JUKEBOXES

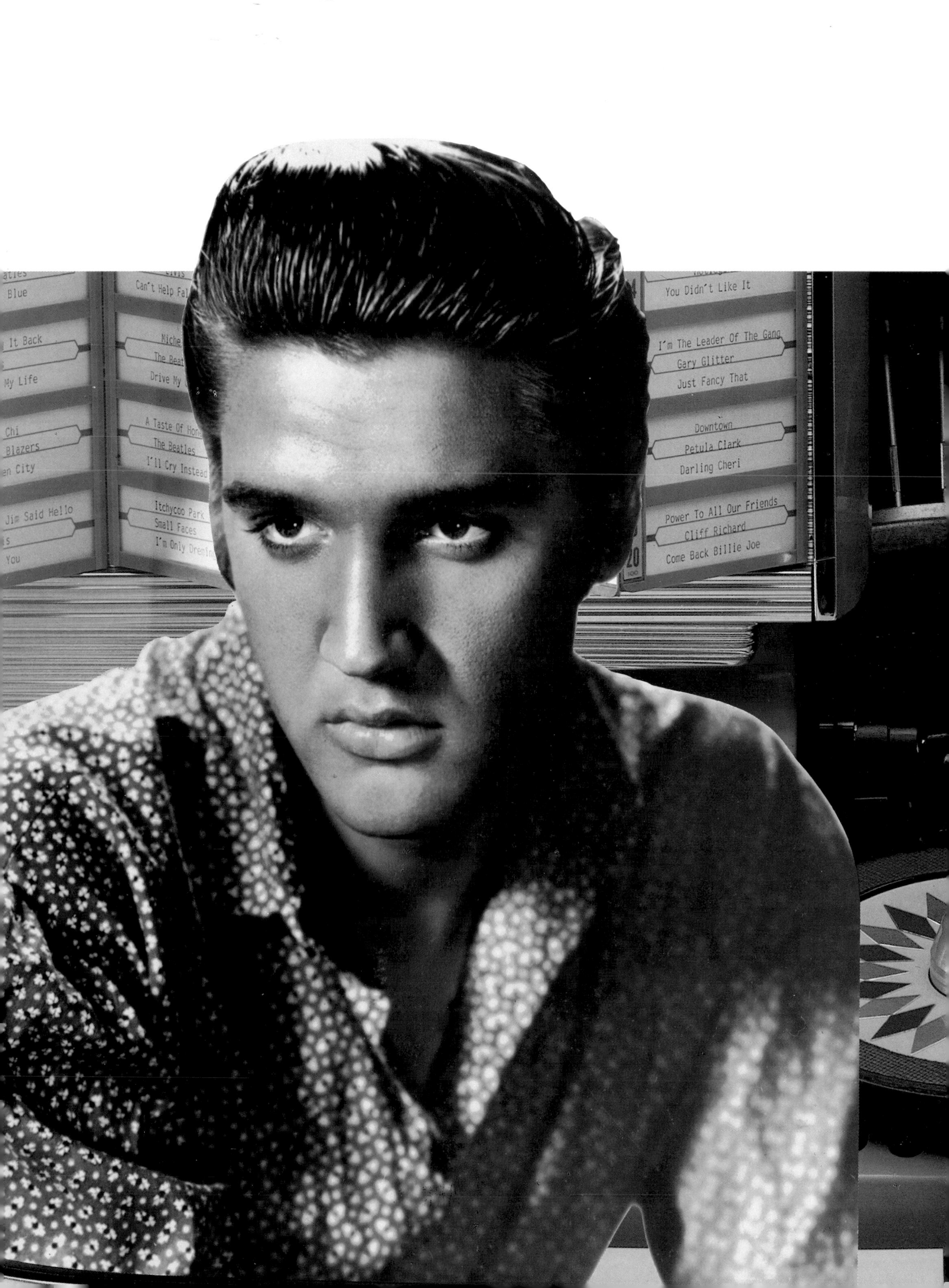

Blue
My Life
Jim Said Hello
The Beatles
I'll Cry Instead
Itchycoo Park
Small Faces
You Didn't Like It
I'm The Leader Of The Gang
Gary Glitter
Just Fancy That
Downtown
Petula Clark
Darling Cheri
Power To All Our Friends
Cliff Richard
Come Back Billie Joe

Michael Adams • Jürgen Lukas • Thomas Maschke

Schiffer Publishing Ltd
77 Lower Valley Road, Atglen, PA 19310

Translated by Dr. Edward Force,
Central Connecticut State University

Photography by Michael Adams, Ullstein Picture Service, Wurlitzer of Germany, Thomas Maschke. The illustrations for pages 113-123 were prepared by Joseph-Wilhelm Reutter of Recklinghausen.

Printed in the United States of America

ISBN: 0-88740-876-1

Library of Congress Cataloging-in-Publication Data

Adams, Michael.
[Musikboxen. English]
Jukeboxes/Adams-Lukas-Maschke.
p. cm.
Includes bibliographical references and index.
ISBN 0-88740-876-1 (hardcover)
1. Jukeboxes. I. Lukas, Jürgen. II. Maschke, Tom.
III.Title.
ML1081.A3313 1996
621.389'33--dc20 95-45129
CIP
MN

Published by Schiffer Publishing, Ltd.
77 Lower Valley Road
Atglen, PA 19310
Please write for a free catalog.
This book may be purchased from the publisher.
Please include $2.95 postage.
Try your bookstore first.

We are interested in hearing from authors
with book ideas on related subjects.

Contents

The Jukebox in the Fifties

The zenith of the jukebox, the era when it embodied a very special significance, can surely be seen in the fifties, not only in America, but also in Europe. Previously it was a coin-operated automat, a machine into which one put money in order to get something. In this case, music. At the beginning of the century, since the medium was new, it attracted a lot of attention, but only curiosity linked the people with the machine. In the fifties, though, the jukebox took on a particular significance: it was the instrument that a young generation used in order to hear its music. Discotheques in our sense did not exist; the market of young people as a potential seller's market had not yet been discovered, and thus the interest in filling their needs was small.

Rock 'n Roll, Rhythm & Blues—these were found mostly in the jukebox. They were not often played on the radio. Thus the little restaurant in which the "genuine" jukebox was found became a hangout, and the jukebox itself became the focal point of the younger postwar generation's enthusiasm for life.

Compared to most of today's customary musical offerings, jukeboxes had one great advantage: the listeners could seek out what suited them. They did not have something offered that they had to consume whether they wanted to or not.

In the sixties, and later, the jukebox still existed as before, but it no longer had any social significance; it took on the role of providing pleasant background music. And with the years, more and more jukeboxes disappeared from the restaurants. Today one has to look for a long time to find a restaurant where time has somewhat stood still and music still comes out of a box.

Debutante: Wurlitzer automat of 1933; it was the first test model, which did not yet bear the Wurlitzer name.

But there are growing numbers of people who are fascinated by the old jukebox. When the jukebox was enjoying its glory days, real wonderworks were wrought which find enthusiastic fans and collectors again today. This book concerns those jukeboxes from the fifties that expressed a generation's joy of life, became famous along with the 45 rpm single, and that for about ten years had a greater significance than just providing background music.

History of the Jukebox

The market for coin-operated automats can be divided essentially into two large groups: There are machines that sell something and others that want to entertain us. The first category includes drink, cigarette and candy automats. In the other group we find slot machines, games, and—naturally—music automats.

A very early example of a coin automat was already found in 18th-century England. There were tobacco automats set up, so-called "honour boxes," where the person who took tobacco was trusted to put the right amount of money in the box. The newspaper boxes that are found on many street corners in large cities function on precisely the same principle to this day. There too, the person who takes the newspaper is trusted to put in the money for it.

On the other hand, in most coin automats the situation is the other way around. Only when the money is inserted does the machine release the product or give the service. Thus they no longer trust in the honesty of the interested person, but rather he is compelled to trust the machine to function properly. Which, as we all know, is not always the case. Numerous patents have tried to deal with the problem that coin automats have in recognizing the right coin and rejecting the wrong one.

Edison with his phonograph. April 18, 1878

It might be mentioned in passing here that at the beginning of the fifties in America, as well as elsewhere, there was a different type of musical automat that accompanied the sale of certain goods with the sound of a melody. For instance, there were musical cigarette machines that gave its own melody to every brand of cigarettes. If one chose Camel, then one heard the title tune from the film "The third Man."

The Machine Speaks

The human desire to produce sounds, especially music, artificially as an art form is ancient. Even in antique times, among the Greeks, machines were built that were capable of producing artificial music. They were used particularly for cult purposes.

Thomas Alva Edison (1847-1931) can be thanked for the ultimate invention of the "talking machine," the phonograph. This man, who also developed the incandescent light among his many inventions, submitted over 1500 patents during his lifetime. It is interesting in this respect that the development of the long-life incandescent light bulb cost Edison $100,000 before the first light burned satisfactorily.

He also devoted his energy to research on mechanical reproduction of sounds. Although there were already numerous attempts to make talking machines and theories on the subject, it was Edison who was able to build the first functioning machine. And it was his voice, in 1877, that was the very first human voice to be heard from a talking machine. The "phonograph," as Edison named it, consisted of a metal drum nine centimeters in diameter, with its axle resting on two supports. On each side was an adjustable ear trumpet (with a parchment membrane), and a steel needle was attached in the middle, over the drum. A thin sheet of aluminum foil was wrapped around the drum; then the device was activated by a crank, and the steel needle made indentations in the foil according to the movement of the membrane.

Now the cylinder had to be turned back again, and a second membrane had to be attached in place of the first one. When one began to turn the crank, what had just been recorded could be heard. The first sounds ever heard after having been reproduced artificially were the words of a nursery rhyme, sung by Edison himself: "Mary had a little lamb."

In an article for the *North American Review*, which he wrote in the summer of 1878, he described the possible future uses of his "phonograph" as follows:

- Write letters and all kinds of dictation without the help of a stenographer.
- Read acoustic books to the blind without making them do any of the work.
- Give instruction in languages and pronunciation.
- Play recorded music.
- Provide a "family album"—record the words, thoughts, etc., of the members of a family in their own voices, as well as a dying person's last words.
- Provide music for musical machines and toys.
- Let clocks announce the time for lunch or when work is over in natural words.
- Preserve characteristics of speech through the exact reproduction of a type of pronunciation.
- Perform instructional purposes, such as recording a teacher's explanations so that the pupils can refer back to them at any time.
- Provide connections with telephones, in order to record simultaneous messages.

Emil Berliner, the man from Hannover, invented the gramophone and the record.

This list is very thorough for a prognosis that was given just shortly after the invention of the phonograph. Although the sound carriers have changed—instead of the cylinder, the flat record, audio tape and compact disc prevail today—the list is quite complete in and of itself, and describes nearly all of the uses of reproduced music or speech that have been realized today.

To be sure, it was to take about ten years until the phonograph could truly gain acceptance and begin to achieve its goals. At first, after initial enthusiasm and wonderful press reports, it disappeared into the shadows again. Edison sold his original gramophone for only $10,000 and 20% of the profits and turned to new tasks, in particular the development of the electric light.

As of the spring of 1888, though, Edison worked intensively on his phonograph again, and in the next five years he applied for more than sixty new patents for it. The aluminum foil was replaced by a hollow cylinder of prepared wax, the needle acquired two sapphire splinters, one pointed for recording, the other blunt for playing.

Edison looked for a process by which the wax cylinder could be copied, so that the buyer of the phonograph could gradually build up a supply of wax cylinders with good recordings. Edison was finally able to solve this problem by hanging the original cylinder in a vacuum and covering it with gold dust, thus making it an electric conductor that could be galvanized. Then it was possible to prepare a master from which copies could be made.

His improved phonograph was also not powered by hand any more, but driven by an electric motor. In practice, this proved to be a wrong decision. The motors were much too big and cumbersome, and they required the availability of electricity, which was not at all to be taken for granted at that time. In addition, the motors were very likely to malfunction. For that reason Edison decided, some time later to be sure, to use a spring-driven apparatus. Since the wax cylinders were still too impractical to be sold, the inventor tried using paper rollers, which turned out to be unsuitable. He was only able to develop a good, inexpensive and easily produced cylinder when he encountered stearic acid-soda salt, which he then had manufactured in great quantities for years and used almost exclusively for the production of his cylinders. The story is told that the improved phonograph, which then began its triumphal journey around the world, was even used in Tibet as a sort of modern prayer wheel, in order to repeat the utterance of prayers.

The Platter that Stores Sound

Edison's phonograph always differed essentially in one way from the recording of sound as it is done on the records that we know today: Loud and soft, high and deep tones are basically represented by deeper or shallower grooves. Thus the sapphire needle moves up and down, touches microscopic hills and valleys, and translates them into audible sounds. Today, on the other hand, this information is not presented by heights or depths in the grooves, but by lateral movements of the cutter during recording and of the needle during playing.

It was May 16, 1888, when Emil Berliner, born in Hannover, Germany on May 20, 1851, introduced a new invention to the members of the Franklin Institute in Philadelphia. He called it the "Grammophon." The essential innovation that differed from Edison's phonograph consisted of the recording no longer being done on a cylinder, but on a round zinc plate. Berliner also did not let his needle swing up and down, but more or less far to the right or left in a horizontal groove.

Thus the phonograph record was born, even though at first it was only a very primitive platter with a rusty surface, which was hardened after recording was completed so that it could be played back again.

The next development of the phonograph record after the "rusty platter" was a zinc disc that was covered with a very thin layer of wax. The needle, attached to a membrane, carved a spiral line of sound movements into this layer of wax, uncovering the bare metal. These opened-up lines could now be etched into the zinc plate in an acid bath, and after the removal of the wax, the finished record was ready to be played.

Above: Berliner's coffee-mill gramophone of 1894; below, Berliner's work force in Washington, 1897.

Page 10: Automatic music boxes of the 19th Century, patent drawings of:

1. G. A. Brachhausen; his music automat offered a selection.

2. Schaub's "Musical Box" had a particularly sturdy mechanism.

3. Parr's "Musical Box" was patented on March 15, 1887.

4. Lochmann's music box was built in Leipzig.

Since the problem of making copies arose here, Berliner continued his research and finally had the decisive idea of using a solid wax platter instead of a zinc plate. The needle could now carve the sound grooves very easily. In addition, the wax platter was treated with graphite, making it an electric conductor, and thus a galvanic counterpart of the original platter could be made of copper. The result was a positive, since the grooves appeared as elevations. From this "master," the platters could then be pressed directly.

Thus Berliner had succeeded in developing two new features that were of decisive importance for sound recording and playing: the lateral movement of the playing needle and the development of the copper master made possible a considerable improvement in sound quality.

As the material for the pressing of records, Berliner utilized a substance that consisted of some 70% finely powdered stone and 30% shellac (a resin), plus additives. The ground stone gave the records the necessary hardness (at that time the needle was run with a very great amount of weight on it), but also made it very breakable.

Thus Berliner had developed all the techniques that were necessary to the production of great quantities of records. The shellac platter, developed by Berliner in 1895 from the substance described above, dominated the market for almost sixty years. It was, and still is, played at 78 revolutions per minute.

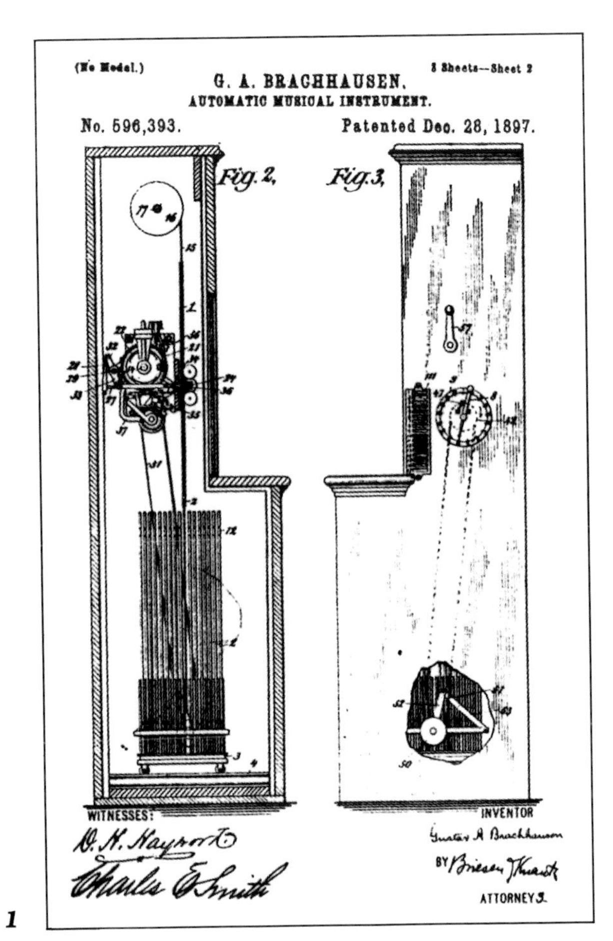

1

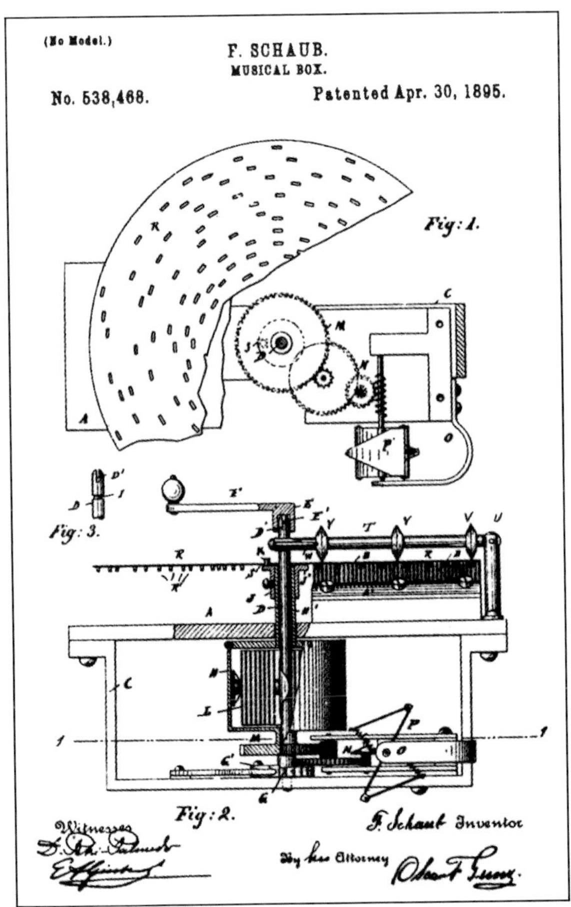

2

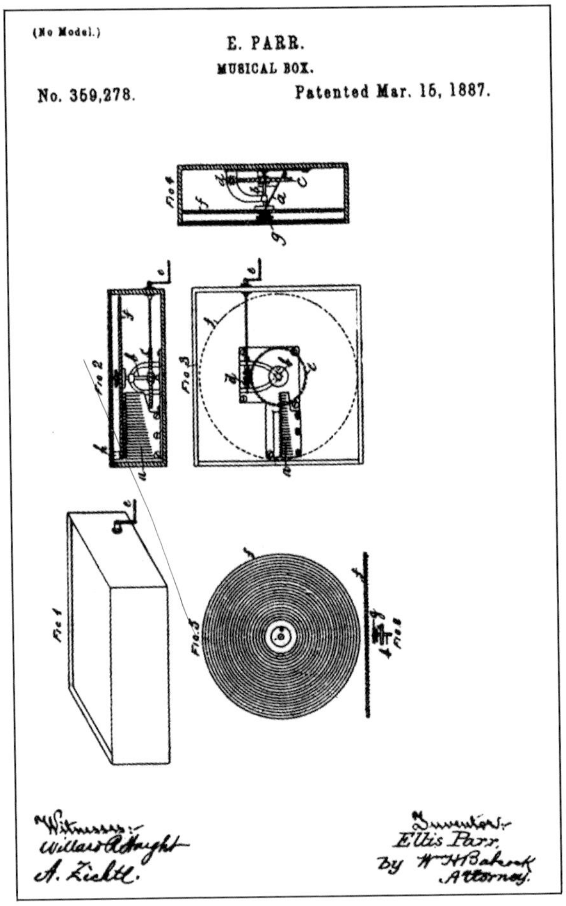

(No Model.)

E. PARR.

MUSICAL BOX.

No. 359,278. Patented Mar. 15, 1887.

Witnesses:
Willard A. Haight
A. Zichtl.

Inventor:
Ellis Parr,
by W^m H Babcock
Attorney.

3

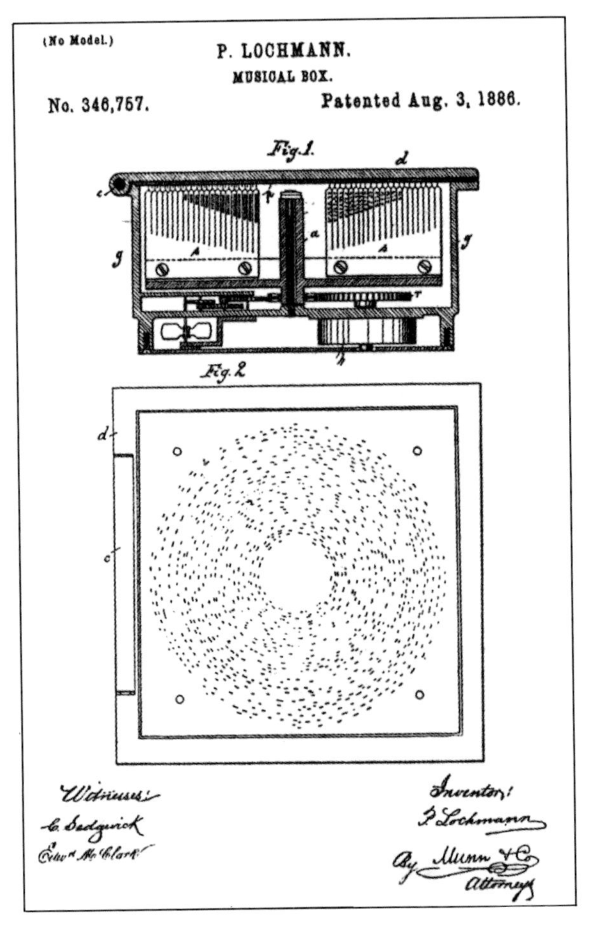

4

The Music Plays Longer

The next development consisted of lengthening the playing time of the record. Around 1900 a platter with a diameter of 30 centimeters still had a recording capacity of only about 4.5 minutes. One means of solving this problem was to increase the diameter of the record. In 1905 a record with a diameter of 50 cm was introduced, but it was unhandy and fragile. Other solutions consisted of placing the sound grooves closer together or reducing the revolving speed. For a long time, both possibilities were followed up and utilized.

Once again it was Thomas Alva Edison, still an enthusiastic proponent of the cylindrical record, who developed the first long-playing record. In 1912 he introduced the first records that had a playing time of forty minutes. This included both sides of the platter; each side had a playing time of about 20 minutes. In the process, Edison retained his principle of varying groove depth, but placed the grooves very close together: he put sixteen grooves within one millimeter. This called for a very resistant material, which he found in the hardenable, fully synthetic resin called bakelite, which the Belgian-American chemist Leo Hendrik Baekeland (1863-1844) had invented.

A particularly hard material was necessary, of course, as the grooves were very fine, thus not very deep, and the knifelike diamond of the needle had to run with a weight of more than 300 grams on it in order to produce sufficient sound volume. Edison's long-playing record (80 rpm) therefore became worn out very quickly.

On September 17, 1931, the first long-playing record intended for the public, with a speed of 33 1/3 revolutions per minute, was finally introduced at the Savoy Plaza Hotel in New York. Records with this comparatively slow revolving speed had already been used several years before that for film sound tracks, with the sound being played from a record parallel to the showing of the film. This first synchronization process helped the sound film establish itself. The "record for everyone" was put on the market on September 17, as noted above, by the American firm of RCA-Victor, along with the record player built for it. To be sure, the record player, with built-in radio, cost between $250 and almost $1000; this high price was presumably the reason why further spread of this innovation did not take place for the time being.

The long-playing record as we know it today was essentially developed by Hungarian-born Peter Goldmark (1906-1977). He used a new record material, a hardenable plastic of the polyvinylchloride family, which has a considerably finer surface structure than the coarse shellac platter. With it, the distance between the individual grooves could be decreased from 0.3 to 0.1 mm. In addition, the groove width could be decreased from 0.13 to 0.07 mm. The sound receiver was now electrified (electrical amplification of signals), and therefore the pressure of the needle could be reduced to 1/100 of what Edison had found necessary several decades before.

In 1766, James Cox of London built a music box; the golden salon piece was 15 inches high and also included a clock and moon-phase indication.

The new records had several advantages:

- They were unbreakable.
- The fine structure of the material and the mixing-in of soot strongly reduced the friction of the needle.
- The length of play, now over twenty minutes, allowed most pieces of music—including symphonies and musicals—to be placed on one record.

The long-playing record developed by Goldmark possessed nearly all the qualities to be found in modern-day records. It was introduced at a trade fair in Atlantic City, New Jersey, on June 21, 1948.

It is interesting in this respect that this new type of record at first found acceptance in Germany only very slowly. As late as 1954, the old 78 rpm shellac records amounted to over 77% of all record production. The production of 33 1/3 rpm records amounted to only 6.2%, with 45 rpm records making up the rest.

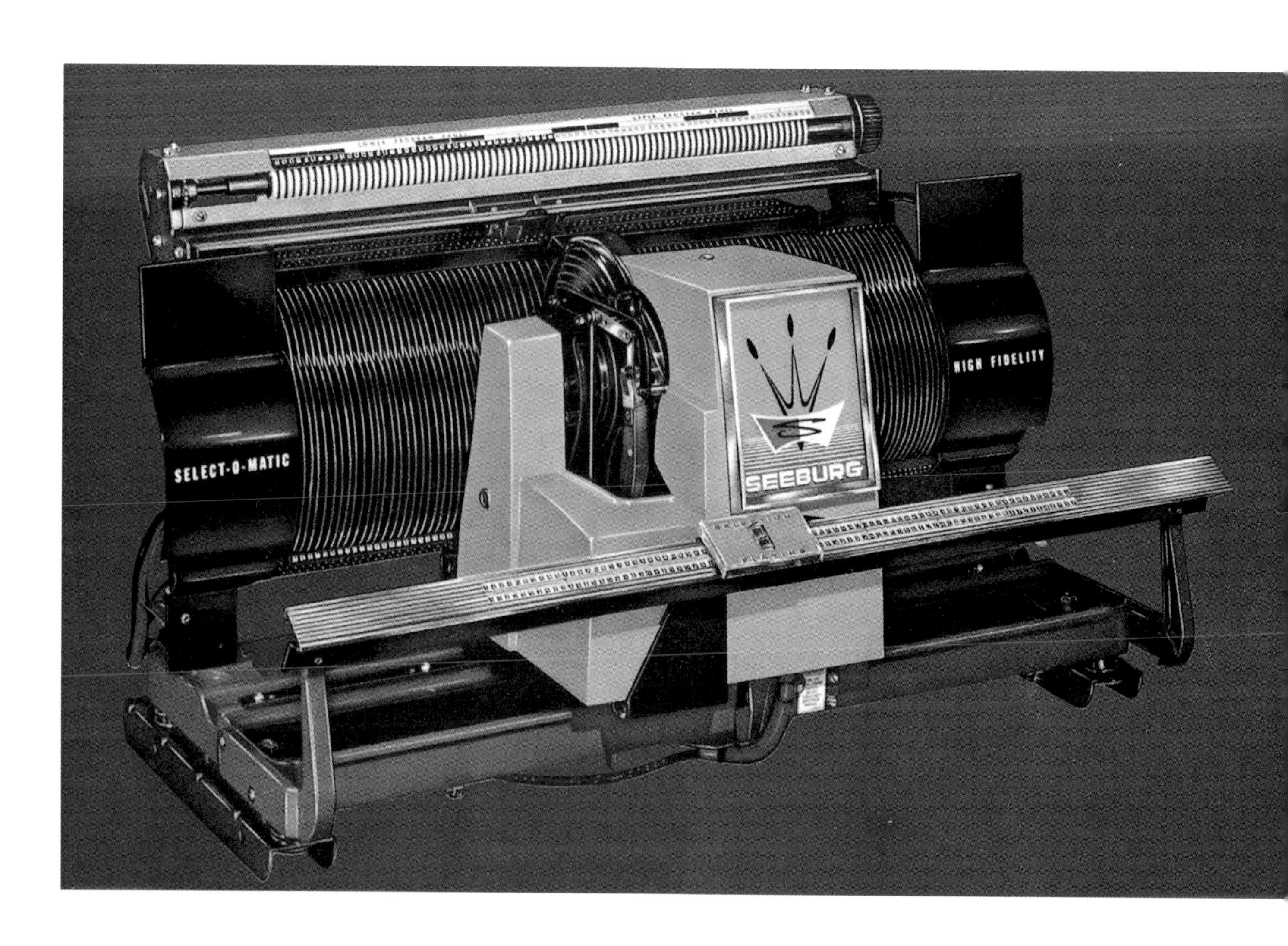

The 45 rpm single was the record of the jukebox! The playing mechanisms of the jukeboxes were adapted to use this size. In the Seeburg mechanism (Select-O-Matic), the records stood vertically, even when being played; in the AMI-i-200 the records stood in a semicircle and were moved out to be played by a selection bolt.

More Speed

When the record firm of Columbia introduced its 33 1/3 rpm LP in 1948, it took a great risk, for everything was oriented to the 78 rpm record, and thus there was a great resistance to changes among the other firms. But the LP could throw several advantages into the bargain. Along with its sturdiness, the weight of the needle arm had also been reduced for the sake of the new microgrooves, and this in turn had greatly improved the quality of the sound.

The firm's great rival, RCA, was "caught cold" by Columbia's venture. In the middle of 1949 RCA came onto the market with a smaller record that had a somewhat higher speed—45 rpm. With that, the 78 rpm had seen its last days. In any case, since more and more manufacturers accepted the 33 1/3 rpm record as made by Columbia, RCA could not help but follow the trend and likewise produce long-playing records.

In spite of that, the firm decided to stick with the 45 rpm record, and in 1950 alone it spent five million dollars on advertising for this product. RCA was of the opinion that the 45 rpm single was better suited to popular music than the LP, which had after all been developed chiefly for longer pieces such as classical music.

In this viewpoint it was supported by one of the most powerful figures in the music industry, Karl T. McKelvy, the general sales manager of Seeburg. McKelvy was well aware of how the single record was ideally suited to jukeboxes. It was small, unbreakable, and light as well. And it had the advantage of being finished quickly. From the company's point of view, the shorter playing time per record meant more business per hour. Since the Andrew mechanism,

In the play "Ballad in Texas" by Kolander, Ingeborg Riehl sang the "Song to the Jukebox," in which the box was praised as the silent consoler.

which Seeburg used for record choice, could also be adapted to the new records relatively simply, Seeburg made the offer to its users that they could exchange their 78 rpm "Select-O-Matic" mechanisms for those made for the 45 rpm singles. Thus the Seeburg firm, which was by now the leading manufacturer of jukeboxes, was able to make a considerable contribution to the eventual success of RCA and the single through its massive support of the 45 rpm single favored by RCA. The other record producers could not hold out long against this; in time, they too brought out their popular music on singles.

Seeburg's Model M 100 B, put on the market in 1950, was the first jukebox for singles. And because of the delivery situation for 45 rpm records, which was still difficult at that time, Seeburg even offered to obtain the records for its customers if they could not obtain them elsewhere.

The other firms such as Rock-Ola, Wurlitzer and AMI were at first not so convinced of the success of the single. Each of their models could be ordered with an optional adaptor kit, which allowed either 45 or 33 1/3 rpm records to be played. Only in 1953 did they change completely to the 45 rpm single.

Thus for a time Seeburg had two advantages that the others could not easily oppose: a selecting mechanism for 100 singles and the decision in favor of the 45 rpm record, which made expensive and laborious adaptations unnecessary.

As can be concluded from a contemporary press announcement, the "Philips Tongesellschaft" introduced this new record at the big German Radio, Phonograph and Television Exposition at Düsseldorf in 1953 under the name of "Philips Minigroove 45." In the report it was further noted that it was a nearly unbreakable record with a diameter of only 17 centimeters, that could be played at 45 rpm. Quote: "In its reproductive quality, the Philips M 45 record equals the well-known 33 1/3 rpm long-playing records. It weighs only 40 grams (the normal record weighed about 200 grams) and, unlike the normal record (78 rpm, with a central hole 9 mm in diameter), has a central hole of 38 mm.."

Automatic Music

Along with the artificial recording of music, there came the possibility of reproducing music as often as desired in two senses: For one thing, once recorded, the piece of music could be played again and again, But above all, it was also possible to copy these recordings; theoretically, as many cylinders, or later platters, as wanted could be made from one recording. Though music could be experienced only "live" up to then, it was now possible for the first time to play the music, once it

was recorded, at any time in any place on earth.

It was easy to take this new medium that attracted so much interest and make a business out of it. The first coin-operated music box was set up at the Palais Royal in San Francisco on November 23, 1889. Louis Glass, Director of the Pacific Phonograph Company, took a rebuilt electric-powered Edison phonograph and equipped it with four sound pipes. Each pipe had its own music slot, the machine could be heard by only one person at a time, and one play cost twenty cents.

The first music boxed suffered from a limited choice of musical material. In 1905 John C. Dunton introduced his "Multiphone." In this mahogany box two meters high, with the shape of an oversize lyre, 24 songs recorded on Edison records could be selected. Great success, though, was not to be attained by the Multiphone; the firm went bankrupt in 1908.

The first automatic music box in the present-day sense was the "Automatic Entertainer" introduced by John Garble in 1906. This automatic entertainer had a pre-selector apparatus and flat records instead of cylinders. The switching mechanism of the device was over a meter in diameter, and the whole machine, glassed on three sides, was 1.50 meters high. But this machine too was not destined to attain success, and in 1908 Garble ceased production. Why this music box did not find acceptance can probably be explained chiefly by the fact that it had miserable sound quality. The sound was tinny, without volume, and disturbed by operating noises.

Only a good twenty years later, in 1927, did the first fully electrically powered devices come on the market. Then there also began the ascendancy of such illustrious firms as Wurlitzer, Seeburg, and Rock-Ola. The invention of the triode and the resulting improvements in sound recording and recreating helped the music machines attain success. The newly developed electrostatic speaker also played a role. Mass production began in the mid-thirties. Now more importance was given to the appearance, with glittering chromed decorative parts, transparent plastic and colored lights making the jukebox more and more attractive.

Another significant impetus was the distribution of jukeboxes by gangster syndicates, who soon became aware that they could make good money out of them. After the era of prohibition (the banning of alcoholic beverages) came to an end and liquor smuggling and blackmarketeering no longer brought in money, such gangsters as Al Capone founded companies that supplied jukeboxes. Transactions with tavern-keepers were similar to protection-money extortion: the gangsters visited the taverns and nightclubs and made it clear to the proprietor that he should acquire a jukebox. Otherwise he could expect violence within the next few days. Thus supported by brute force, the manufacturers had sold more than 300,000 jukeboxes in the USA by the time World War II began. Most of them were controlled by gangster syndicates.

Page 14:
An advertisement from the fifties.

Below:
How an artist conceived a modern bar of the fifties (drawing from a sales brochure).

Platten- die Geld bringen!
(Dave Mackersie,
Hammondorgel und Pfeifsolo)
St. Louis Blues / Basin' Street Blues
(Clyde McCoy –
»Der Schluckauf-Trompeter«)
C 15 802 · CF 15 802
Heideröslein / Friesenmädel
(Golgowsky-Quartett und Will Glahé)
F 43 637 · D 17 637
Schwedenmädel (Schwedische Rhapsodie)
Johnny
(Lys Assia mit Orchester)
F 43 696 · D 17 696
O My Papa (O mein Papa)
Another Dawn, Another Day
(Ray Anthony, Solotrompete)
C 80 377 · CF 80 377
Rosen erblühten, als wir uns fanden
Sei lieb zu mir
(Gitta Lind mit Orchester)
A 11 491 · U 45 491
Der Mann am Klavier
Keine Frau ist so schön wie die Freiheit
(Golgowsky-Quartett)
F 43 694 · D 17 694
Bella Ida
Das Schönste vom Tag ist der Morgen
(Die Drei Travellers)
F 43 752 · D 17 752
Schwedische Rhapsodie / Jamaican Rumba
(Mantovani und sein Orchester)
F 43 666 · D 17 666
Sugar Blues Tear It Down
(Clyde McCoy –
»Der Schluckauf-Trompeter«)
C 80 276 · CF 80 276
Nicolo, Nicolo, Nicolino
Die Frau kommt direkt aus Spanien
(Fred Weyrich u. Orch. Béla Sanders)
A 11 568 · U 45 568
Ich hab' Heimweh nach St. Pauli
Zum letzten Mal, Kapitän
(Rodgers Duo)
A 11 569 · U 45 569
Amorada / Delicado
(Svend-Asmussen-Quintett)
F 43 603 · D 17 603
F 43 327 · D 17 327
Gespielt und Gepfiffen (Pack Fast)
A-O-Rag
(Dave Mackersie, Hammondorgel und Pfeifsolo)
F 43 585 · D 17 585
Marching Strings / Ecstasy
(Edmundo Ros und sein Orchester)
F 43 587 · D17 587
Bella Ida
Das Schönste vom Tag ist der Morgen
(Die Drei Travellers)
F 43 752 · D 17 752
Bon soir, bon soir
In Paris, in der Rue Madeleine
(Vico Torriani mit Orchester)
F 43 650 · D 17 650
Amorada / Delicado
(Svend-Asmussen-Quintett)
F 43 603 · D 17 603
Bella Bimba / Auf Jamaica schenken
abends die Matrosen
(Sylvia Dahl m. Orch.)
A 11 549 · U 45 549
Istanbul / Maple Leaf Rag
(Joe Fingers Carr)
C 2665 · CF 2665
Moulin Rouge Thema / Agnes-Walzer
(Mantovani und sein Orchester)
F 43 561 · D 17 561
Wo der Wildbach rauscht
In 1000 Meter Höh'
(Golgowsky-Quartett)
F 43 691 · D 17 691
Jodeln kann ich nur, wenn ich verliebt bin
Tango der Nacht
(Vico Torriani mit Orchester)
F 43 614 · D 17 614
Diesmal muß es Liebe sein
Am Ende der Welt
(Gisela Griffel mit Orchester)
A 11 469 · U 45 469
Sugar Blues Boogie / Hells Bells
(Clyde McCoy –
»Der Schluckauf-Trompeter«)
C 80 376 · CF 45 80376
Kostenlos
erhalten Sie den
MUSIK-BOX-DIENST
sowie alle
Neuerscheinungslisten
u. Bezugsquellen-
Nachweis durch
TELDEC
SCHALLPLATTEN-
GESELLSCHAFT MBH.
HAMBURG 36
Hohe Bleichen 22
TELE FUNKEN
TELEFUNKEN · DECCA · CAPITOL · SCHALLPLATTEN

The Big Names

Until the beginning of the fifties, the history of the automatic music box, particularly the jukebox, took place almost exclusively in the United States. The origin of the "big four" companies will be portrayed here.

Seeburg

As they so often do behind the great names in the realm of music boxes, one person stands behind the name of Seeburg: Justus P. Seeburg (originally Sjoebert). Born in Sweden in 1871, he emigrated to America in 1886 to seek his fortune in the New World. After first having worked in several piano factories, he decided in 1907 to become independent and founded the J.P. Seeburg Piano Company.

There he produced his first coin automat, the "Orchestrion," a reworked piano with mechanically played violins, mandolins, flutes, triangles and percussion instruments was operated mechanically by electric bellows. With the growing popularity of the phonograph, the Orchestrion was taken out of production in 1927.

Seeburg turned to music boxes, and between 1927 and 1934 he produced the "Audiophon" series, the changing mechanism of which required a lot of space and made the machine very wide. On the other hand, they were equipped with the latest electric amplifiers and dynamic speakers, so that they sold well.

As so often in later years, it was the Seeburg firm that, in 1938, created a turning point in the design of jukeboxes, by adding lights to them. The "Symphanola Classic," introduced in 1938, attracted tremendous attention. For the first time, this model featured transparent plastic panels behind which low-power light bulbs were located, and the jukebox began to glow. It was not long before Wurlitzer produced models that, in terms of design and illumination, exceeded the Seeburg models.

Seeburg then devoted his main emphasis to offering his clientele something that the others did not have. In 1939 the "Wall-O-matic" was introduced. Small and space-saving, it could be set up practically anywhere. This small control box was connected to the actual music machine by a cable. Thanks to the multiple cable (one line per possible selection, 20 titles), the "Wall-O-matic" could be placed directly at the guests' tables, which was done in the better restaurants.

In contrast to the other manufacturers, particularly Wurlitzer, the design philosophy of the Seeburg firm was rather restrained. New models were not a departure from the well-known trends, but rather were based on time-tested juke-box designs. Thus the four models available in 1940 (Cadet, Colonel, Commander and Concert Master) were further developments of 1939 "Classic" and "Vogue" models. And there were good reasons for that, for the acoustic characteristics of the case were very essential for the extremely good sound reproduction for which the Seeburg models were always renowned. The Seeburg models were solid; in these monumental machines, nothing stuttered or flickered, and they showed no exciting play of colors. But they had the best sound.

Left:
Records that bring in money. This 1954 advertisement is likewise a list of the hit records of that year.

Next two pages:
Advertisement by Seeburg and its distributor, in "Automatenmarkt," May 1, 1954.

Ein großer Wurf

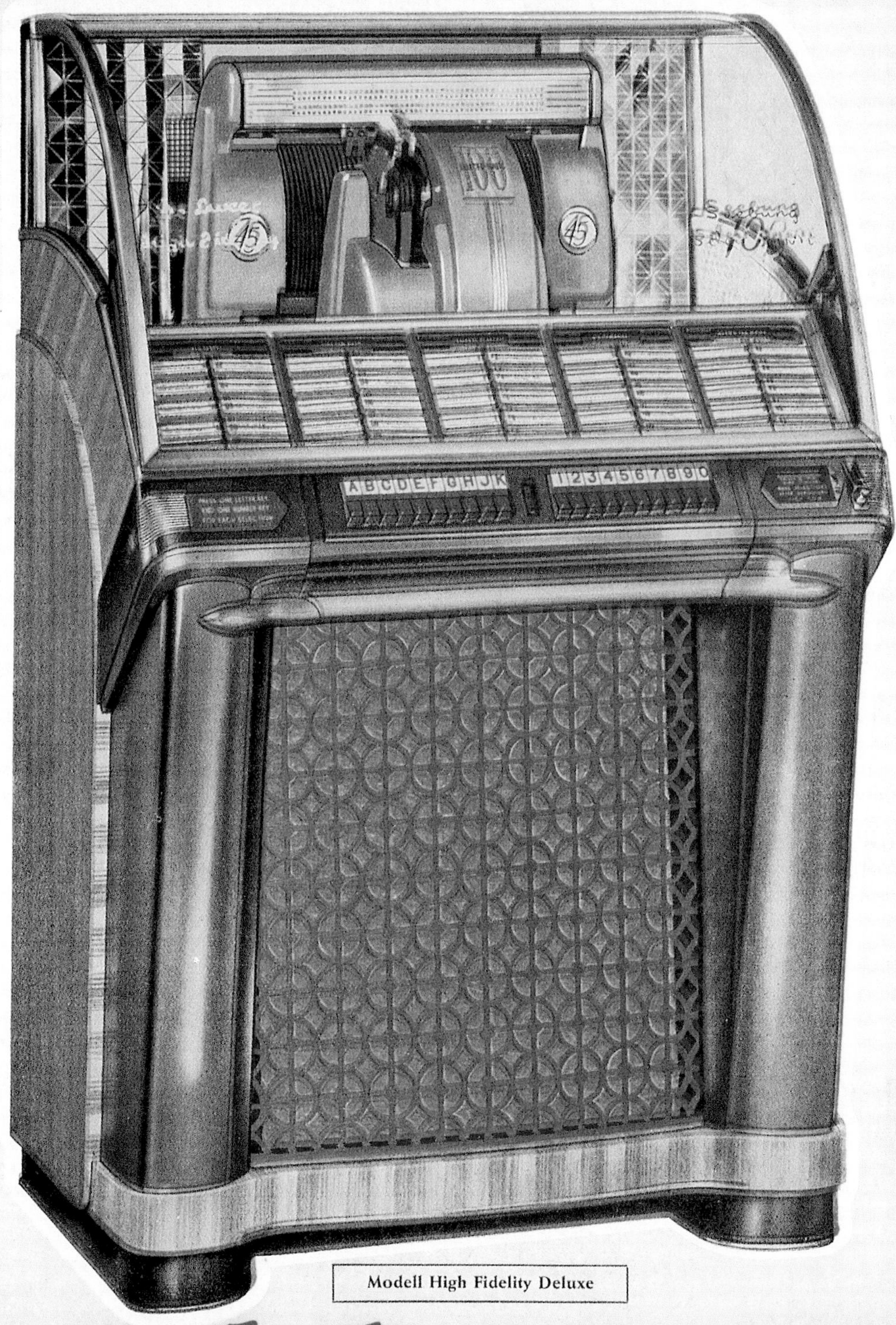

Modell High Fidelity Deluxe

ist glücklich gelungen

SEEBURG

H-100-FG Deluxe *und* M-100-W

as sind die Spitzen-Musikautomaten, auf die der deutsche Aufsteller so lange gewartet hat.

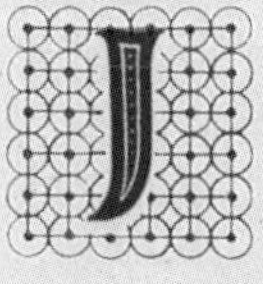

eder SEEBURG-Käufer ist zufrieden! Diese großartigen Boxen empfehlen sich von selbst weiter.

an kann deshalb nur sagen: Kauf einer SEEBURG ist eine Geldanlage ohne Risiko.

Bedienungsanweisung in deutscher Sprache · Einwurf für 10-, 50-Pf.- u. 1-DM-Stücke

GENERALVERTRETUNG FÜR DIE BUNDESREPUBLIK:

LÖWEN-AUTOMATEN

ZENTRALE: Braunschweig, Berliner Straße 5/6 · Sammelruf 21616

NIEDERLASSUNGEN:

WEST: Düsseldorf 10, Tussmannstraße 74, Fernruf 491411/12

NORD: Hamburg 6, Merkurstraße 38, Fernruf 430693

Wurlitzer
Hi-Fidelity
POPULAR AND CURRENT HITS
WESTERNS AND FOLK TUNES
POLKAS AND SPECIALTIES
CLASSICAL AND OLD FAVORITES
A B C D

In American advertisements for the Wurlitzer firm, the following sentence turned up now and then: "A name famous in music for over two hundred years." That may sound somewhat strange at first, for the history of music boxes is, after all, not that old. But in fact it is true, for the ancestors of the Wurlitzer family, who came from Silback, Saxony, settled in England and earned a living producing sounds since 1659. In 1853, Rudolph Wurlitzer emigrated to the United States and held several modest jobs until, in 1880, he began to produce pianos. He was helped by his younger brother Anton.

The firm's actual climb into the world of music automats began in 1908, when Wurlitzer purchased the de Kleist factory, which produced carnival equipment and organs. These organs were used to produce musical entertainment during silent movies, and as the motion picture began its triumphal journey, Wurlitzer also won fame with the "Wurlitzer Motion Picture Orchestra." And it was this "Mighty Wurlitzer" that accompanied the heroes of silent movies since 1910 and won the Wurlitzer firm its great fame.

When Rudolph Wurlitzer died in 1914, his three sons took over the firm.

With the emergence of sound movies toward the end of the twenties, the market for the "Mighty Wurlitzer" collapsed. Then the meeting of Homer Capehart with Fanny Wurlitzer became a stroke of luck. Capehart, who knew the music automat market very well, brought not only his thorough-going knowledge but also a usable machine. It was made by a small company called Simplex, which had developed a very good mechanism for changing records and had designed a case for it. Fanny Wurlitzer had the capital and could make good use of it.

Thus Wurlitzer succeeded in outdoing its major competitors—AMI, Seeburg and Mills—very quickly. In 1934 the firm's complete production numbered about 15,000 pieces. Then, when Wurlitzer more or less rolled up the market, they were able to sell at least 63,000 jukeboxes by the end of 1936.

In that year, Wurlitzer had record sales of 44,397 jukeboxes, and for the first time since the difficulties of the previous years, the firm was in the black again. Purely for business reasons, old jukeboxes that had seen years of service were taken in trade by Wurlitzer and then destroyed. By the beginning of World War II, the sale of jukeboxes by Wurlitzer remained around 30,000 per year.

Up to that time, jukeboxes had been more or less smooth cabinets with attention-attracting fronts. This changed with the introduction in 1940 of Wurlitzer's Model 700, the sides of which were also decorated with "rich Italian onyx." This onyx, though, was nothing more than a very good plastic imitation, behind which optional lighting could be installed.

In the same year as the 700, namely 1940, a very similar but larger model, the 800, was introduced. It was identical to the 700 in many details. The notable new feature was the front grille, which included three illuminated columns in which bubbles rose. At that time it attracted great attention, especially when the columns were illuminated from the back. In the middle of 1941, Wurlitzer introduced the 750 E, its first jukebox in which the selection mechanism was electrically controlled.

At that time, Wurlitzer controlled over 50% of the entire market. The public, though, believed that Wurlitzer had 70 to 80% of the market.

Left:
From a Wurlitzer brochure:
Operation of the Wurlitzer carousel jukebox, which came on the market in 1954.

Next two pages:
Miss Symphonie advertised for Bergmann & Co. and their jukebox with 80 selections. She was the ideal model of an elegant lady of the time.

The Wurlitzer 2500 was offered in four versions (ready for Hi-Fi stereo!) with 104 and 200 selections. Wurlitzer stated in the advertising for this box: "To portray the new Wurlitzer visually, the printer had a truly insoluble problem: in the customary multicolor printing, it was impossible to reproduce realistically the soft light effects of the 2500 models in their fine nuances and gradations."

Miss
Symphonie

WURLITZER MODEL SELECTIONS
2504 — 104

The hyphen in the name "Rock-Ola" was probably present because of the fact that the Canadian David C. Rockola became annoyed when people mispronounced his name, turning it to "Rocko" or "Rockla" instead of Rockola. With the hyphen between the syllables, it was sure to be pronounced right.

He made his first commercial success with scales at the age of 23, after emigrating to the United States. When Prohibition was repealed, bars were opened everywhere, and the people wanted moderately priced entertainment. Rockola thought in terms of music. But when it became known that his firm also wanted to develop jukeboxes, Wurlitzer saw its position endangered and instituted legal action against Rockola on account of patent infringements, with punitive damages of one million dollars requested. Rockola immediately instituted a countersuit for two million dollars. This case dragged on over several years and cost Rockola half a million dollars.

Of the "Big Four" (Wurlitzer, Seeburg, Rock-Ola and AMI), Rock-Ola was the firm that got into the juke-box business last. But it was just the right time to succeed in business, and yet late enough to avoid repeating the mistakes that the other competitors had made in the bitter years of the Depression.

As can also be seen later with IMA-AMI, AMI always had a taste for unusual yet practical business connections. Much like the Danish firm of Jensen, which later became IMA-AMI, the National Piano Manufacturing Company which was founded in 1909 had a subsidiary firm, the National Automatic Music Company, which bought all its products and then sold them. In 1925 the two firms became one and became a branch of the Automatic Musical Instrument (AMI) Company.

In 1926, AMI began to produce automatic music boxes, which featured what was then a very advanced selection mechanism. In the prewar years, AMI introduced a series of elegant jukeboxes, of which the "Singing Tower" was their top-line model. Around 1940, AMI turned almost completely away from the original idea of the jukebox, favoring the "Automatic Hostess" telephone system instead. This system, first introduced in 1939, was actually very simple: The "music box" was connected by a coin microphone with a central station, in which disc jockeys operated whole series of record players. This system had one definite advantage: the choices were theoretically unlimited.

The music requested by the listener was then played into the building. AMI advertised it with the explanation that one was dealing with people, with the friendly voice of a hostess. This was supposed to be "as close as one of your own family and as distant as an angel in heaven."

With "Mystic Music" in 1941, Rock-Ola placed a very similar system on the market, but regarded it less as an alternative to the jukebox than as an extension. Conventional jukeboxes were modified so they could serve as "Mystic Music" stations, though they did not work instantly, as a lever had to be pushed first.

With the outbreak of World War II, though, these systems disappeared, since the "hostesses with the tongues of angels" had other things to do: They worked at the telephone switchboards of the military or of industry.

Right:
Advertisements for Rock-Ola (1940, Rock-Ola's Luxury Lightup Phonograph), Rock-Ola "Strike up the band" (1940), and "smallest bar-box ever," the Model 332, and for the Wurlitzer wall box.

Pages 26-27:
"The promised surprise has succeeded. Perfected in technology, in appearance and in quality of reproduction, is our Rock-Ola wall box..." (original advertisement, 1960).

"AMI-i-200 is the midpoint—it attracts attention with light, style and sound!" (Original AMI advertisement, 1958).

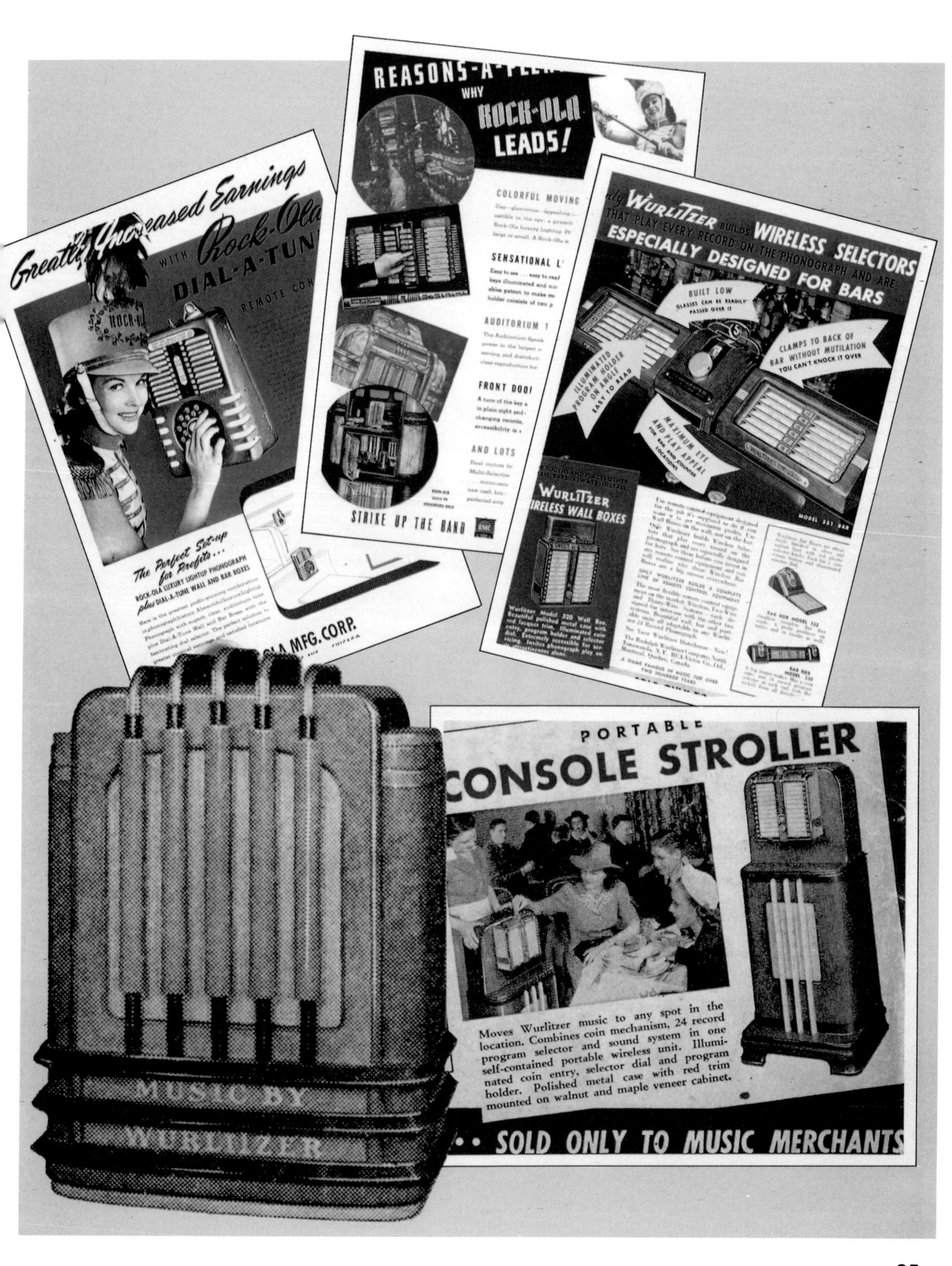
Greatly Increased Earnings
WITH Rock-Ola
DIAL-A-TUNE
The Perfect Set-up for Profits...
ROCK-OLA LUXURY LIGHTUP PHONOGRAPH
plus DIAL-A-TUNE WALL AND BAR BOXES
OLA MFG. CORP.
REASONS-A-PLE
WHY
ROCK-OLA
LEADS!
COLORFUL MOVING
SENSATIONAL L
AUDITORIUM T
FRONT DOO
AND LOTS
STRIKE UP THE BAND
WURLITZER BUILDS WIRELESS SELECTORS
THAT PLAY EVERY RECORD ON THE PHONOGRAPH AND ARE
ESPECIALLY DESIGNED FOR BARS
BUILT LOW
GLASSES CAN BE READILY PASSED OVER IT
CLAMPS TO BACK OF BAR WITHOUT MUTILATION
YOU CAN'T KNOCK IT OVER
ILLUMINATED PROGRAM HOLDER ON ANGLE
EASY TO READ
MAXIMUM EYE AND PLAY APPEAL
WURLITZER
WIRELESS WALL BOXES
PORTABLE
CONSOLE STROLLER
Moves Wurlitzer music to any spot in the location. Combines coin mechanism, 24 record program selector and sound system in one self-contained portable wireless unit. Illuminated coin entry, selector dial and program holder. Polished metal case with red trim mounted on walnut and maple veneer cabinet.
• • • SOLD ONLY TO MUSIC MERCHANTS
MUSIC BY
WURLITZER

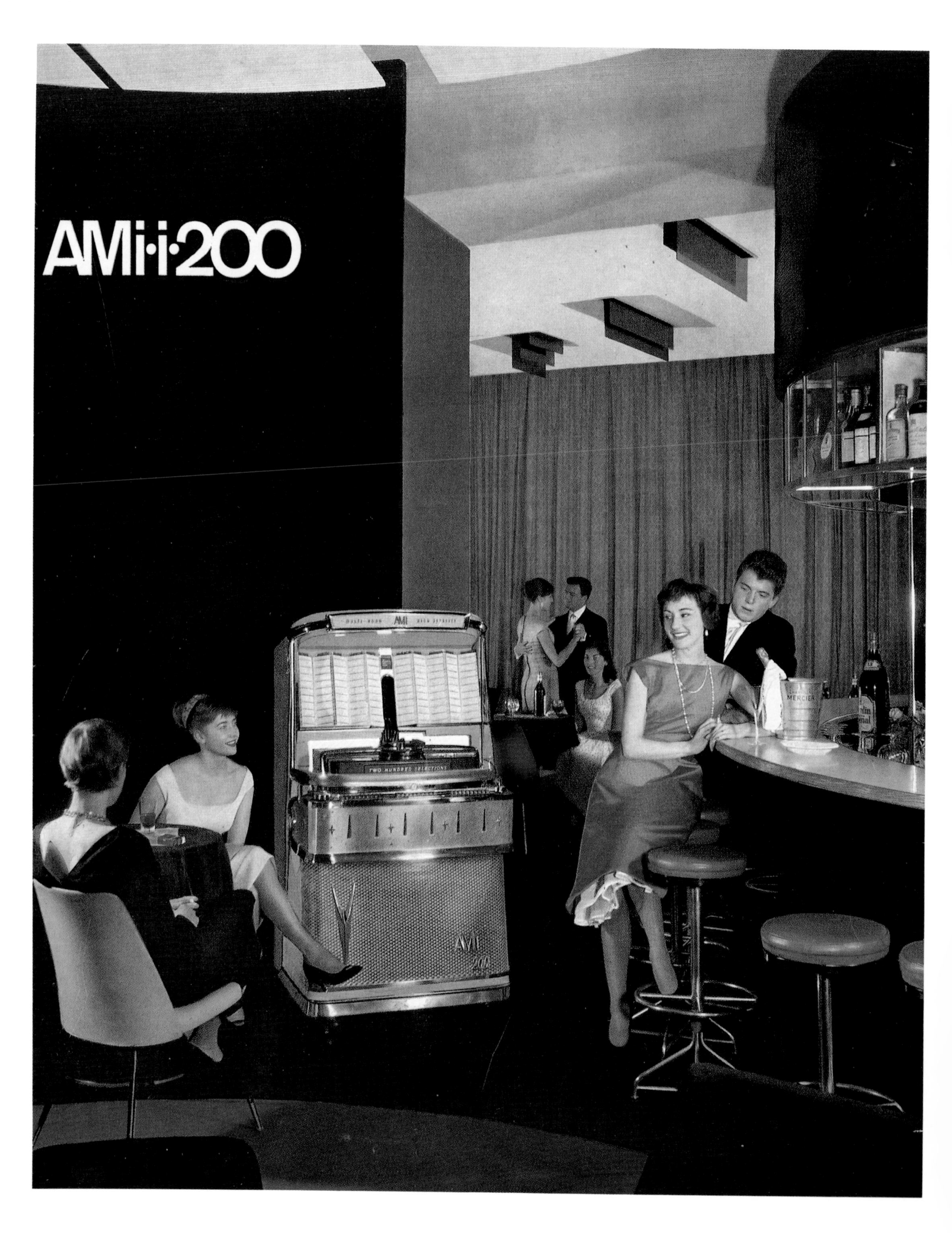
AMi·i·200
TWO HUNDRED SELECTIONS
MERCIER

ROCK-OLA WANDBOX

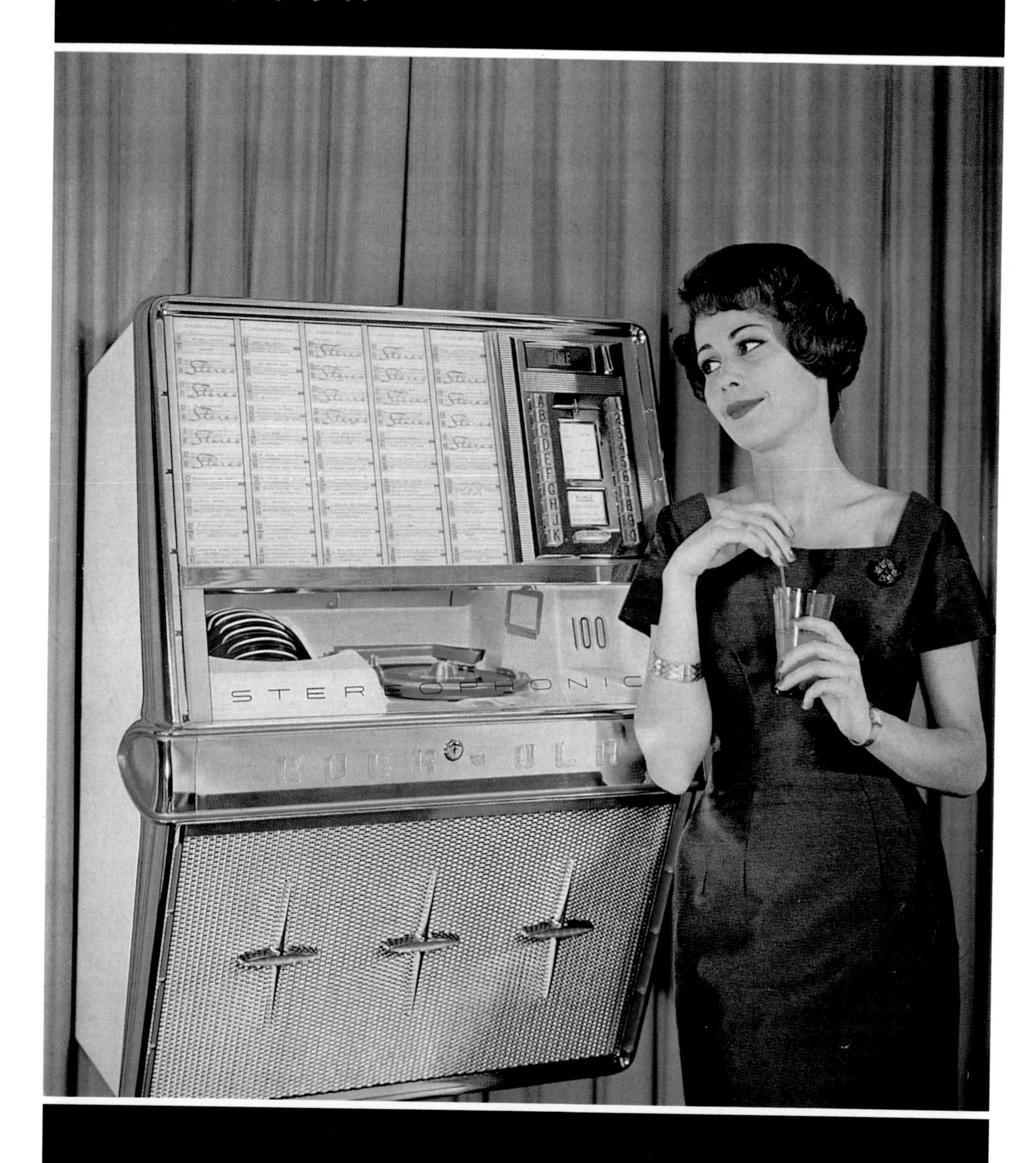

The Wurlitzer 1100 excited the mood in the forties.

The War Years

For the American juke-box industry, World War II was a disaster in which production came to a complete standstill. In December 1941, the government forbade the production of coin automats as of February 1 of the following year, and also informed the manufacturers that they should reduce their production by 75% as of the same day.

Jukeboxes were at the very top of the list of non-essential goods. In the spring of 1942, no music automats were built any more. The firms were occupied with the production of essential goods for the war; Rock-Ola produced guns and ammunition boxes—and did so on three eight-hour shifts. Wurlitzer busied itself producing parts for aircraft and other war materials. On account of these events, restructuring naturally took place within the firms. At Seeburg, the developmental department had been increased from 70 to 130 technicians, and they continued to produce armaments after the war. Of course, that was not necessarily profitable, but it allowed Seeburg to carry out tests that would not have been possible otherwise. Ironically, the market for jukeboxes boomed in the war years, when the production of jukeboxes had been stopped. Very long work hours and poor working conditions were nothing unusual in view of war production. In the little free time that they had, people wanted to amuse themselves. In addition, a good number of jukeboxes were shipped to the soldiers overseas to bring up their morale. In short, these times were a gold mine for the sellers. Since there were no new jukeboxes, the old machines were hauled out of the cellars and put back into service. Old "P 10" and "Multi-selector" were back at work, along with thousands of the "Universal" cabinet and the Wurlitzer "Victory."

During World War II, the more important resources such as metals were needed by the government and therefore not released for use in other manufacturing. Thus it was difficult to produce jukeboxes. Since many materials were already rationed in 1941, Paul Fuller, for example, had to use a wooden tone-arm mount in the Wurlitzer Model 950, because metal had become scarce. And during the war, Wurlitzer—just like the other manufacturers—simply no longer had any of the materials needed to make jukeboxes. Metal, plastic, work force and record-changing mechanisms were all lacking.

For that reason, Wurlitzer had cabinets for their jukeboxes made of wood. "Victory" was a universal rebuilding kit, with which old jukeboxes could be updated. According to estimates, Wurlitzer was able to sell 18,000 of these Victory kits.

With the end of World War II, the juke-box industry had reasons for great optimism. In America, the product was more popular than ever before, and it had also reached Europe with the GIs, so that a new market opened up there. After the production limitations had been abolished, the juke-box manufacturers began to compete to place new machines on the market. The machines used by the consumers were five or more years old, and the desire for new models was correspondingly great.

On to the Fifties

Seeburg

After the war, Seeburg was the first manufacturer to put a new model on the market: the P 146. With the 146/48 series, Seeburg had a reasonably priced, reliable and unusual-looking jukebox. The last two digits in the number referred to the year of production. All of them were equipped with the reliable Freborg 20-fold selection mechanism, and the title listings were located inside the selection buttons.

At the same time, though, this Freborg mechanism was a major disadvantage to the Seeburg models, for it was carried in a steel frame, so that the record changer was not visible. On the other hand, the models made by Wurlitzer, Rock-Ola and AMI had record changers in which the moving parts were readily visible, and this in turn was a strong sales tool. Although the P 146 sold well at first, since there was nothing else available, Seeburg soon had to look for an open mechanism. Once again it was Seeburg who, in 1948, offered the public a new jukebox that was at least three years ahead of everything else that was then produced in the industry. This was the M 100 A. No more curves and arches, no plastic with lights behind it, and no handcrafted wooden case. The M 100A had a rectangular form with sharp angles and with pointed corners, it had chrome decoration and a strong neon light. And although—or perhaps precisely because—this jukebox was so far

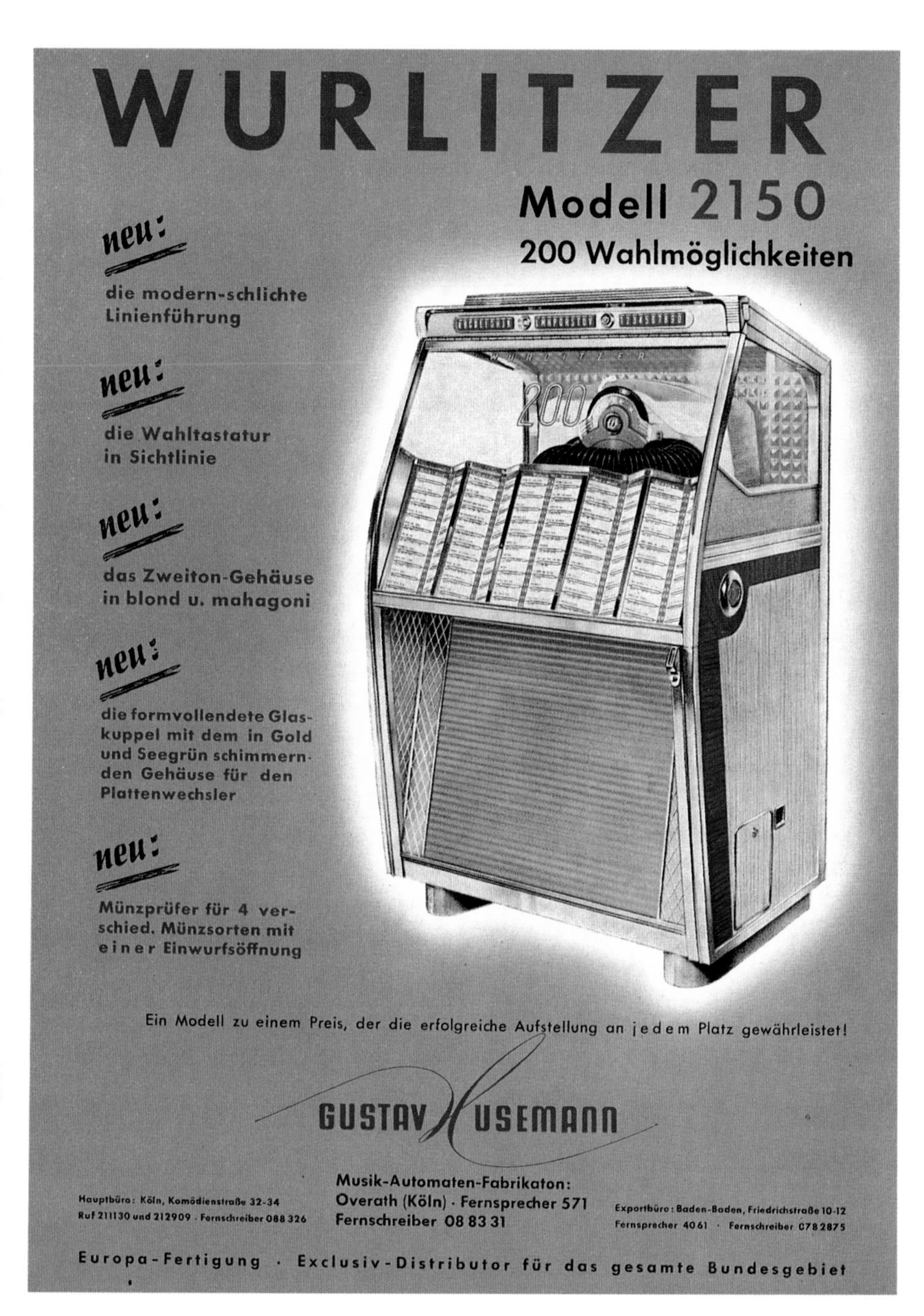

The design of the boxes, like that of the advertisements, has changed: Wurlitzer Model 2150 of 1957, and the "Seeburg remains Seeburg" advertisement.

SEEBURG bleibt SEEBURG...
200
Seeburg
200 Wahlmöglichkeiten
Ideale Tanz-musik
Seeburg
Affettuoso
Affettuoso

ahead of its time, it was also regarded as a sensation by the public. Next to it, all the others looked archaic. The "Select-O-Matic" was seen in the best places; even loyal Wurlitzer and Rock-Ola customers now turned to Seeburg.

In the M 100 A, probably for the first time, the design principle that prevails at the present time appeared: "Form follows function," the shape is subordinate to the function.

The new mechanism of the M 100 A, completely visible, was simply revolutionary—just like the exterior of the box. In the M 100 A, the fifty records are lined up vertically; in front of them is the playing mechanism, mounted in a plastic housing that reveals only the tone arm and the record itself. When a selection is made, this entire playing mechanism travels along the records until it reaches the right one. The record is brought out of the group and placed on the record platform. The tone arm is equipped with two needles, so that it can play either Side A or Side B. After being played, the record is put back into its place, and the next selection can begin. With the M 100 B in October 1950, Seeburg introduced the first jukebox for 45 rpm singles.

In 1953 the 100 W and HF 100 G models appeared. The arched glass panel had become larger and also covered the title listings. And the selection keys were no longer just small buttons, but were similar to the keys of a piano.

The Model 100 W was the lower-priced type, despite which it looked superb. The demands of the times, though, were best met by the Model HF 100 G; it promised High Fidelity. In an age when High Fidelity was a major topic in the private sector, Seeburg could offer its purchasers a model that, as the buyers knew well, promised good sound quality and even adorned itself with the concept of "Deluxe Hi-Fidelity." The Seeburg was advertised as being very similar to a Hi-Fi system for private use.

Once again it was the Seeburg firm that, in 1955, set a milestone in the history of the jukebox with the Model V 200. This jukebox was noteworthy for a number of reasons. It was a very large machine, exceeded only by the Wurlitzer Model 1500. It was the first machine that offered 200 choices. It was also the first to have an electronic storage unit, and the first to offer the possibility of putting in two separate selections (dual system).

And there was something else that made it very special: The V 200 was the last model that was produced under the Seeburg family, which withdrew from the business a year later.

After the great competitors Rock-Ola, Wurlitzer and AMI had succeeded in reaching or surpassing the magic number of 100 selections that Seeburg had once set, Seeburg was once more able to set a new record. The 200 choices, of course, made particular demands on the selection system if it were not to become too big to read. Thus Seeburg decided on a solution that Wurlitzer had already applied in the Model 1100: a revolving title drum. For this, Seeburg used a cylinder on which the title strips were located. The chosen titles were magnetically recorded with the help of the Tormat data storage unit.

Thus the electromechanical method that had been used until then was done away with.

With the Model 222, Seeburg introduced the first stereophonic jukebox, which achieved the stereophonic effect with the help of at least two auxiliary speakers. High frequencies above 380 Hz were broadcast by the auxiliary speakers, while the bass came from the speaker in the box itself. This system was very practical, for the slim box itself was not at all capable of offering the necessary distance between the two speakers that is required for good space orientation. This, though, could be achieved with the auxiliary speakers. It is much harder for the human ear to locate deeper sounds, so that here it was sufficient to have them coming directly out of the box.

With the Seeburg dual system, which was also offered in the Model V 200 (optionally), the records could also be offered at different playing prices. The difference between the higher- and lower- priced songs could be determined when the machine was installed.

Wurlitzer

After the war, Wurlitzer introduced the Model 1015, the best-known and best-loved jukebox of all time. The front of this box is a single lighted arch. The Model 1100 followed in 1947 and was described as "ultramodern." Its light was somewhat more glaring, and the selection knobs had been changed, so the user would have a better feeling of control. Conservative juke-box owners could choose the Model 1080, which was designed in 1946. In their opulent appearance, Wurlitzer models essentially followed the philosophy that the jukebox itself should provide as much pleasure as the music that was played by it.

Within eighteen months, to the end of December 1947, nearly 80,000 jukeboxes could be shipped out of the Wurlitzer factory, comprising over 70% of the total production of this industry. Wurlitzer was floating at the top of this wave of success, but that was not to last long. After the war years, during which the main trend was toward a change of pace, most people were drawn back to home and hearth, with the effect that fewer jukeboxes were sold. All the major manufacturers found themselves in dire straits, for they had all greatly overestimated the future market.

A particular feature first used in the Model 1080 A was the new "Cobra" tone arm. This curved arm, made by the Zenith Corporation, was ideally suited for use in jukeboxes. Until then, very heavy tone arms had been used, with needles that had to be replaced often. They wore out the (shellac) records very quickly. Zenith proclaimed that with their new tone arm, a record could be played 2000 times and still retain 95% of its sound quality.

The Wurlitzer Model 1250 is amazingly similar to the Seeburg M 100; one could almost describe it as a parody. An interesting feature of the 1250, which was also equipped with a second tone arm so that the B sides of the records could also be played, which increased the number of possible selections to 48, is the fact that the protective cover over the selection mechanism is very high. This was done because Wurlitzer considered increasing the number of records to enlarge the possible number of selections.

With the 1400 and 1450 models of 1951, Wurlitzer basically tried to build on the success of the forties. While the Model 1400 was finished in walnut, the special feature of the 1450 was its "textileather." This material was supposed to be resistant to water, alcohol and other damage. And with the Wurlitzer 1450, a jukebox could, for the first time, be ordered in different versions: mahogany, light brown, blue and brown. But the selection mechanism was still the same as in the 1250, and thus could offer the public nothing new. To be sure, Wurlitzer tried to turn this disadvantage into an advantage; the firm advertised the advantages of this tried and true system, and described its competitors' large numbers of selections as not necessary. So Wurlitzer was of the opinion that its record-selecting mechanism, in which the records lay flat, one above the other, was more reliable than those of its competitors. And from questionnaires received from its customers, Wurlitzer could conclude that about 75% of the business was done with only eight to ten titles; the rest of the records were seldom played, slowed the selection process and raised the service costs.

All of that might have been true, but even Wurlitzer's admission of limited choices did not last long. In the very next year, 1952, the Model 1500 was introduced, a jukebox that could play 104 different titles—four

The Wurlitzer factory premises in Tonawanda, New York (from a Wurlitzer advertisement of 1954).

Oh, mein Papa

Die reizende Schweizerin **Lys Assia** hat sich mit dem Lied „Oh mein Papa..." in die Herzen aller deutschen Schlagerfreunde gesungen. Aber auch in den USA ist dieser Schlager ein großer Erfolg. In der „HIT PARADE", der amerikanischen Schlagerparade der Juke-Box-Aufsteller, steht er in fast allen Staaten an erster Stelle.

Die 10 Musik-Box-Spitzenschlager in Bremen

1. Granada (Vico Torriani auf Decca)
2. Annaliese (Hans A. Simon auf Elektrola)
3. Jambalaya (Gerh. Wendland auf Polydor)
4. Sugar Bust (Doris Day auf Columbia)
5. Preußens Gloria (Polydor-Blasorchester auf Polydor)
6. Alte Kameraden (Musikkorps des Bundesgrenzschutzes auf Polydor)
7. Pferdehalfter (Kilima Hawaiians auf Philips)
8. Sugar Blues (auf Telefunken)
9. Olé Guapa (Alfred Hausèr auf Polydor)
10. Sag' doch nicht immer Dicker zu mir (Hans A. Simon auf Polydor)

Lösen die »Perlfischer« das Pferdehalfter ab?

Die Kilima-Hawaiians glauben, daß nach dem „Pferdehalfter" nunmehr ihr Song „Die Perlfischer" (Vaya con Dios) ein großer Erfolgsschlager werden wird.

Jährlich 2 Millionen Schallplatten

14,5 Mill. Langspielplatten mit 45 U/Min. hat die USA-Industrie seit dem Erscheinen dieses Plattentyps auf den Markt gebracht. Insgesamt hofft die US-Schallplatten-Industrie, die Produktion in diesem Jahre um rund 12 v. H. gegenüber 1952 zu steigern und etwa 225 Mill. Stück zu produzieren (1952 = 200 Mill.). Das größte „Plattenjahr" war bisher 1947 mit 250 Mill. Stück Umsatz. (Cash-Box)

Welche Platten sollen in die Juke-Box?

Für den Aufsteller von Musikautomaten ist die genaue Beobachtung der Schlager und ihrer Beliebtheitskurven von großer Bedeutung. Nach diesen Beobachtungen kann er seine Plattenwahl vornehmen. Die richtige Auswahl aber bedeutet für ihn bares Geld

Der AUTOMATEN-MARKT will seinen Lesern bei dieser Arbeit behilflich sein. Nicht jeder wird Zeit und Gelegenheit haben, sich die notwendigen Unterlagen selbst zusammenzuholen. Die Wahl an Hand des Kataloges ist oft unzureichend. — Vorspielen lassen kostet Zeit und — der eigene Geschmack ist nicht immer gleichbedeutend mit dem des Publikums.

Wie in jedem Monat, hat der NWDR auch im Januar einige Tausend seiner

Rudi Schuricke

Hörer nach ihren 10 Lieblingsschlagern befragt. Hier ist das Ergebnis:

1) Annaliesa,
2) Rote Rosen, rote Lippen, roter Wein,
3) Pferdehalfter,
4) Moulin rouge,
5) Soviel Wind und kein Segel,
6) Es blüht eine weiße Lilie,
7) O, mein Papa,
8) Die Glocken von San Marco,
9) Sag doch nicht immer Dicker zu mir,
10) Ich möchte nochmal zwanzig sein.

Interessanterweise hat sich die Auswahl und Reihenfolge gegenüber der Dezember-Umfrage trotz Wechsels des Personenkreises kaum verändert. Dabei weisen nur die ersten vier Schlager eine Stimmenstufung auf, während die letzten sechs praktisch auf gleicher Höhe liegen.

Der gefühlvolle Schlager dominiert also — und bei einer Reihe von Musikautomaten, die wir beobachteten, zeigt sich die gleiche Tendenz — jedenfalls, soweit sie in Lokalen normalen Gepräges aufgestellt sind. „Annaliesa" hat an vielen Plätzen das „Pferdehalfter" endgültig aus seiner Spitzenposition verdrängt, z. T. hält sich dieser Erfolgssong jedoch noch mit fast unglaublicher Zähigkeit.

Im übrigen sind Vico Torriani und Rudi Schurike begehrte Favoriten. Auf einzelnen Plätzen schiebt sich schon mal wieder ein flotter Marsch in die Spitzengruppe mit hinein. Bei unserer stetig größer werdenden Basis wird die Beobachtung der Platten bald viel sicherer Rückschlüsse auf den Publikumsgeschmack zulassen als Repräsentativerhebungen der Sender, die allerdings als Vorschau auch für den Musikautomatenaufsteller ihren Wert behalten werden.

Deutschlandlied Exportschlager

Die „Deutsche Grammophon Gesellschaft", Hannover, sicherte sich mit ihrer Aufnahme vom „Deutschlandlied" mit dem großen Sinfonie-Orchester einen großen Exporterfolg. Wie der NWDR in seiner Sendung „Heidelberger Palette" am 15. Dezember berichtete, hat sich die deutsche Schallplattenindustrie trotz der Auslandsnachfrage noch nicht zur Neuauflage des „Horst-Wessel-Liedes" entschließen können.

Vico Toriani, der derzeitige Liebling der deutschen Weiblichkeit

more than their greatest competitor, Seeburg, could offer with the M 100 A. For this, Wurlitzer had not developed a new selection mechanism, but had simply installed two of them in the new models 1500 and 1550 (the latter with imitation leather). The double tone arm was located in the middle, with the well-known Wurlitzer stacks of records to the right and left of it, with each having been expanded to form two groups of 26 records each. But the Wurlitzer 1500 was not mechanically reliable, and not very successful; only 8383 were produced.

In his book "Vintage Jukeboxes," Christopher Pearce describes this model as a "mechanical nightmare." The whole thing was further complicated by the fact that these models were capable of playing both 78 and 45 rpm records. The Model 1500 could even be modified to play only 33 1/3 rpm LPs, so that it was theoretically capable of playing 26 hours of uninterrupted music without repeating itself once.

In the following year, 1953, Wurlitzer introduced reworked models, the 1500 A and 1150 A, featuring an arched viewing window of exactly the same size as the Seeburg M 100 A, M 100 B and M 1200 C. Here again Wurlitzer utilized the double Simplex mechanism.

Even the successor models of 1953 still "trusted" in the old Simplex mechanism. The 1400 series was replaced by the 1600 and 1650 models, with the 1600 being available optionally at any speed, while the 1650 was only built for 45 rpm records. A little reworking resulted in the 1600 A and 1650 A models in 1954, which were also offered as "Hi-Fidelity" versions—as 1600 AF and 1650 AF. These were the last Wurlitzer models that used the Simplex mechanism.

Left:
Which records go into the jukebox? This page from the "Automatenmarkt" of February 1954 is a real historical document.

Wurlitzer too could finally offer a new mechanism. Similar to the Rock-Ola models, it was a drum in which the records were arranged upright in a circle. But unlike the Rock-Ola, this drum was horizontal. The mechanism was more complex than the Rock-Ola type, and it also looked somewhat more cluttered. But that was hidden, so in the eye of the beholder the new Wurlitzer mechanism appeared elegant and simple.

When a selection was made, one of two arms, depending on the side chosen, grasped the record and moved it into the vertical playing position. The record platform was covered by a circular arch, to which the tone arm was mounted. When the record was in position, the motor began to turn, the tone arm was lowered and glided along the grooves of the record, apparently in defiance of all the laws of gravity.

And although the new Model 1700 had essentially taken over the case of the 1600 series, that was not very important. The attraction was the new mechanism. With it, Wurlitzer once again became a serious competitor in the juke-box business.

With the Wurlitzer "Centennial Model 2000" of 1956, the firm brought out a model that showed such clear resemblance to automobile construction as had never yet been seen at Wurlitzer. The standard "Centennial" 1900 model offered the usual 104 possible selections. The star, though, was the Model 2000, with which Wurlitzer for the first time could offer a jukebox with 200 titles. Originally the 200 title labels were placed on separate pages, each one with two titles, which could be looked through like a book. This was operated by a motor and controlled by push-buttons. Unfortunately, this system was so prone to breakdowns that modifying kits were soon offered, turning the movable title listings into fixed form.

At the end of the sixties, Wurlitzer produced machines that were beneath its standards and did not live up to the customers' requirements. The Wurlitzer firm was not prepared to invest large sums in its juke-box department either. As long as it supported itself, it was to be maintained. In 1972, with the Model 1050, a last desperate attempt was made. The 1050 was introduced to fit into the trend toward nostalgia and earlier styles like Art Deco. It was not meant to be a copy, but "a modern interpretation" of the legendary 1015, and this model does indeed awaken nostalgic memories.

But even the 1050 could not bring on the hoped-for upturn, so the American Wurlitzer firm made its withdrawal from the juke-box business at the end of 1973.

In 1960 Wurlitzer opened a branch in Germany which exists to this day. The "Lyrik" was the first model built there.

AMI

Shortly after the war, this firm came out with the Model "A" (Mother of Plastic), probably the most lavish model that was ever produced by a juke-box manufacturer. This "monster," 1.75 meters tall, was a wonderful seller. The AMI "B" that came on the market in 1947 was somewhat smaller and did not have the jeweled decoration of the Model "A," so as to please the more conservative public.

While the other juke-box manufacturers were still having it out with questions like the number of selections and the speeds, AMI was able to enlarge its share of the market steadily. The firm always had one of the best-developed record changers, although it was the oldest model. AMI had the great advantage that its changing mechanism, developed as early as 1930, could play both 78 and 45 rpm records. Thus no high developmental costs for a new mechanism were necessary.

In the fifties, the AMI boxes did not at first stand out particularly in terms of eye-catching designs. The Model F, introduced in 1955, nevertheless had a few special features to offer, such as its "Multihorn Hi-Fidelity" and "Sonoramic" sound. The smallest change, though, was to prove to be the one that helped the model win its success: the AMI F was available in various colors. With the Model G of 1956, AMI could even add to this. For the first time in the firm's history, a completely new changing mechanism, the X 200, was introduced, offering 200 selections. The X 200 functioned very similarly to the Rock-Ola system. And on May 5, 1958, the first AMI I 200 produced in Europe was built in The Hague.

IMA-AMI

Music boxes of the IMA-AMI type were manufactured by the Jensen Musikautomates A/S firm, then one of the largest musical goods producers in Europe. The firm was founded in 1950 and grew quickly. This was originally a result of the fact that Jensen had sought out the right trade partners. Thus it had found in the Oskar Siesbye A/S a financially strong trade and finance company that handled the exporting of its products.

Even more interesting, though, was its collaboration with a second firm, the Dansk Grammofon Automat A/S. This firm bought all the machinery produced by Jensen that was retailed in Denmark. The Dansk Grammofon Automat A/S then took charge of the locations, the servicing, the purchasing of records, and all the other things that were connected with the juke-box business. All that the Jensen Musikautomates S/A had to do was build the jukeboxes.

In 1953 the American firm of AMI became aware of Jensen, and they began to work on a merger that subsequently came to pass. The Jensen firm became closely connected to the AMI firm and received a license to build Danish versions of the AMI jukeboxes. These boxes produced by Jensen were given the brand name of IMA-AMI.

In the middle of 1954, the IMA-AMI firm of Copenhagen introduced its first model, the J120, on the European market. The J120 was made for 45 rpm singles and offered 120 selections. According to the license contract signed by the Danish IMA-AMI and the American AMI Incorporated firms in December 1953, IMA-AMI used original parts made by AMI Incorporated. Thus the J120 was equipped with the AMI 120 playing mechanism and two speakers. The volume control and the tone arm with its special needle were Jensen designs.

In the middle of 1958, the Danish firm of IMA-AMI halted production of jukeboxes, to be able to concentrate completely on the production of television sets.

The fascination of the automat conquered many areas in the fifties. "Nite and Day" supplied many of the necessities of life after the stores had closed, and the people no longer bought their travel tickets at a window, but from an automat.

Rock-Ola

Just like the other manufacturers, Rock-Ola too reworked its prewar program during the recession period of the mid-forties. The Model 1426, introduced in 1947, was based essentially on the Model 1422 but had a metal grille instead of the wooden type, plus several other design differences that departed from the Art Deco style of the 1422. Only the Model 1428 of 1948 was partially a new development.

In order to compete with the high number of selections in the Seeburg models, in 1950 Rock-Ola introduced the "Rocket," which could play both sides of a record. The Rocket offered 50 selections and, although it was essentially a 78 rpm player, could be modified for 33 1/3 or 45 rpm records. Since it was difficult to develop a completely new mechanism quickly, Rock-Ola had decided for the time being to use a tone arm with two heads, one above and one below the record. When the underside of a record was selected (odd numbers), then the record was pulled out of the stack in the usual way, but the turning direction of the motor was reversed and the lower side of the record was played. In this way, Rock-Ola was able to double the possible selections. It must be added that the "Rocket" was a jukebox that required constant adjustment.

In 1952, Rock-Ola was able to offer some competition to Seeburg with the "Fireball" (Model 1436). This model had a carousel mechanism for 120 45 rpm singles. The problem of housing such a large number of records (sixty in all) as compactly as possible had been solved in an elegant manner: The records were housed in a vertically rotating record drum. When the record had reached the selection position, a stopping mechanism was released and it could be taken out and played. A gripping arm pulled the record out and placed it, depending on whether side A or B had been selected, appropriately on the turntable. This mechanism in its basic details was then adopted by AMI.

With this model too, Rock-Ola could tell its buyers that the open mechanism with the rotating drum would be such a great attraction for the guests as they put in one coin after another to watch the fascinating operation of the mechanism. And naturally it was also said that Rock-Ola had now succeeded in exceeding Seeburg in the number of selections provided by offering 120 instead of Seeburg's 100. But in the fifties, Seeburg dominated the market, particularly in terms of the design of the jukebox. Rock-Ola's "Comet Fireball" (Model 1438) was strongly based, at least externally, on the Seeburg 100 C.

Right:
Of course the market was dominated by American jukeboxes, but German products like the "Diplomat" by Wiegandt could also do well.

Jukeboxes From Germany

The jukebox became known in Germany through the American forces: the GIs, far away from home though they were, did not want to do without their music and their jukeboxes. At first, though, German interest was quite meager. For the most part, used jukeboxes were imported from America. But it did not take long until the American manufacturers discovered the possibility of a European market and set up sales agencies in Germany and other European countries.

Then the juke-box industry really blossomed in Germany. In 1953, jukeboxes were already dominating the scene at the Frankfurt Automat Fair. At this fair, for the first time, the brand-new Seeburg HFG could be seen; with this introduction, the importing of factory-new Seeburg jukeboxes began. Until then, they had only been available in Germany used or from older series. With Seeburg, too, the last of the leading American juke-box manufacturers had decided to establish an agency in Germany.

It is noteworthy that in the year of 1953 three German jukeboxes were already on display: the Wiegandtbox (40 title selections), the Tonomat (100 selections), and a prototype of the Symphony, a 40-selection box made by Bergmann of Hamburg. The first two were intended for 78 rpm records. At this time, though, the American manufacturers were all producing almost exclusively boxes for the 45 rpm singles.

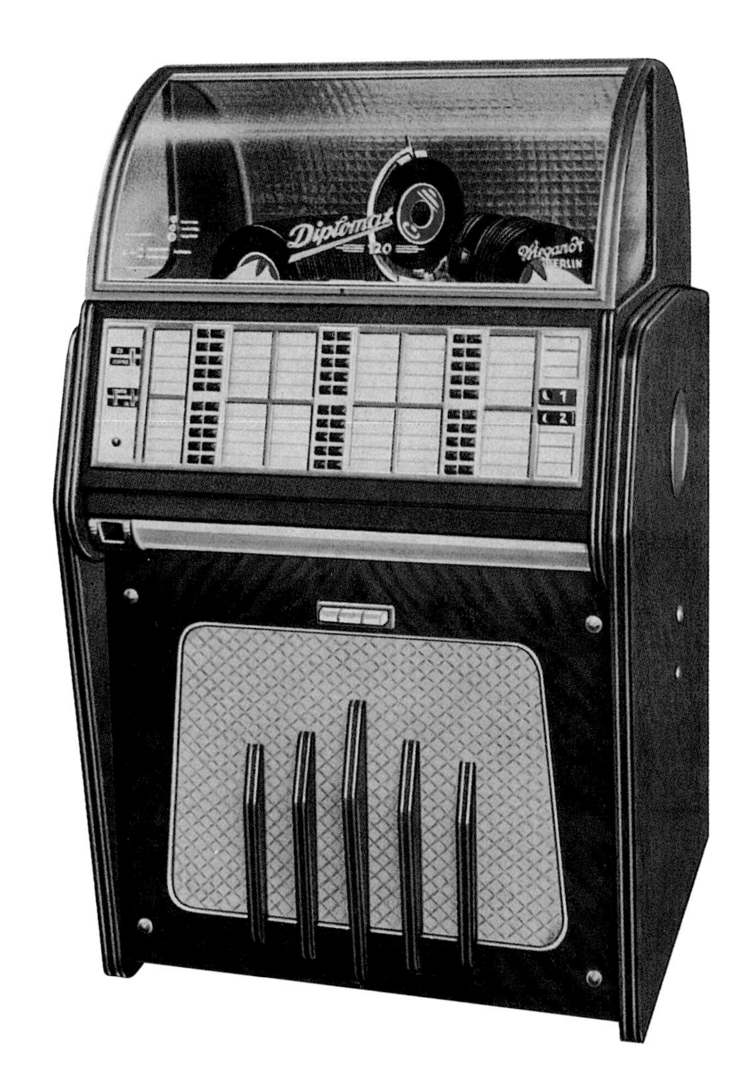

Wiegandt

The autumn fair of 1952 in Frankfurt already showed the first jukebox by a German manufacturer. It was the firm of Wiegandt in Berlin that ventured onto the market with a design of its own. The Wiegandtbox, with its forty selections, was the forerunner of the popular Diplomat boxes. In 1954, the Diplomat 120 for 45 rpm records and the Wiegandt jukebox, with forty selections (78 rpm), were first shown at the Berlin Industrial Fair. The Diplomat had a bird's-eye maple finish with plenty of chrome, offered color effects and an arched glass cover that spanned the entire selection mechanism; the records stood in a semicircle on a rack behind the playing mechanism. With 120 selection buttons—one for every record side—the operating panel took up a lot of space. In 1957 the Diplomat 120 was reworked and introduced with a new case. Like the Tonmaster, it was equipped with a volume control operable from outside.

The Diplomat C 120 Stereo, introduced in 1958, was Wiegandt's first stereo jukebox, and had also been modified strongly externally. With its stereo tone arm, an auxiliary amplifier for the second channel, and an added speaker, the electric features could handle stereo. Unlike its predecessor, this jukebox had a combined letter and number selection system with full pre-selection capability.

Tonomat-Automaten of Neu-Isenburg

More and more German firms ventured to take part in the juke-box business. The second of them was Tonomat, a newly opened factory especially for jukeboxes, which introduced its "V 102" in 1953. This box was continually developed further, from the "Telematic 100" of 1955 through the "Telematic 200" (1957) to the "Panoramic" of 1958, and finally the "Teleramic 200," the first big German jukebox.

The Telematic 200 was a further development of the V 102 and T 100 boxes, which included the telephone selection disc typical of Tonomat jukeboxes, located in the middle of the box. Selections were made with three numbers: the choice of the top side of a record always began with "1" and that of the flip side with "2." The plexiglas covering of the top was curved inward, while a frosted glass panel in the lower part of the box was illuminated from inside.

The carousel mechanism was also retained for the Telematic, though the record magazine took on a larger diameter so it could hold more records. After being lifted, the record was centered on the vertical turntable by an electromagnetically controlled plate, and after playing, the tone arm automatically returned to its resting place. Only the lifting mechanism and the pick-up brush were mechanically controlled. The T 200 offered the possibility of being operated at differing playing prices; up to 40 long-playing records could be selected by the user instead of singles and paid for at 10 pfennig more.

The Panoramic 200, introduced in 1959, not only found success in Germany but also was exported in large numbers. It was also available with full stereo as the "Panoramic 200 S." The Tonomat firm tried to give this box an "American" look. It was the first German jukebox to be equipped with a popularity scale.

Bergmann

One year later than Tonomat, in 1954, the firm of Th. Bergmann became the third German company to introduce a jukebox: the "Symphony

40." It was followed the next year b the improved "Symphony 80," whic according to the manufacturer be came the best-selling jukebox mad in Germany. This box, complete produced in Germany, corresponde to German taste with its simple, wel shaped design. It had more in com mon with a music cabinet than wi the heavily chromed American juke boxes of that time. The records wer stored vertically in a magazine, an the selection was made by pus buttons. First one of four groups wa chosen, then the number of th record, from 1 to 20, was indicated The playing process could be ob served through a window, and th interior was illuminated by two tubu lar lights. The speaker grille con sisted of overlapping tile-shape glass strips and was illuminated b a ring of light. In 1958 Bergmann in troduced its first large jukebox, th "Symphonie 200."

NSM

The firm of NSM-Apparateba was founded in Braunschweig i October 1952 by G. W. Schultze, H Nack and W. Menke. Schultze an Nack were proprietors of th Löwenautomaten wholesale house while Menke was a designer.

The firm was very successfu with jukeboxes of the Mint series, th names of which still sound quite fa miliar to this day: Triomint Roulomint, and the best-known post war German automat, the Rotamint Further models were the Rotamin Record, the Mint and the Rotamin Luxus and Rotamint Super models

This interesting excerpt come from an NSM press release: "Lik everything new, both inside and out side this country, the introductory process of the Fanfare naturally has taken its time. Meanwhile, in regard to the internal German market, it car be regarded as finished. The com parison with the introduction of ou successful Rotamint in its day is

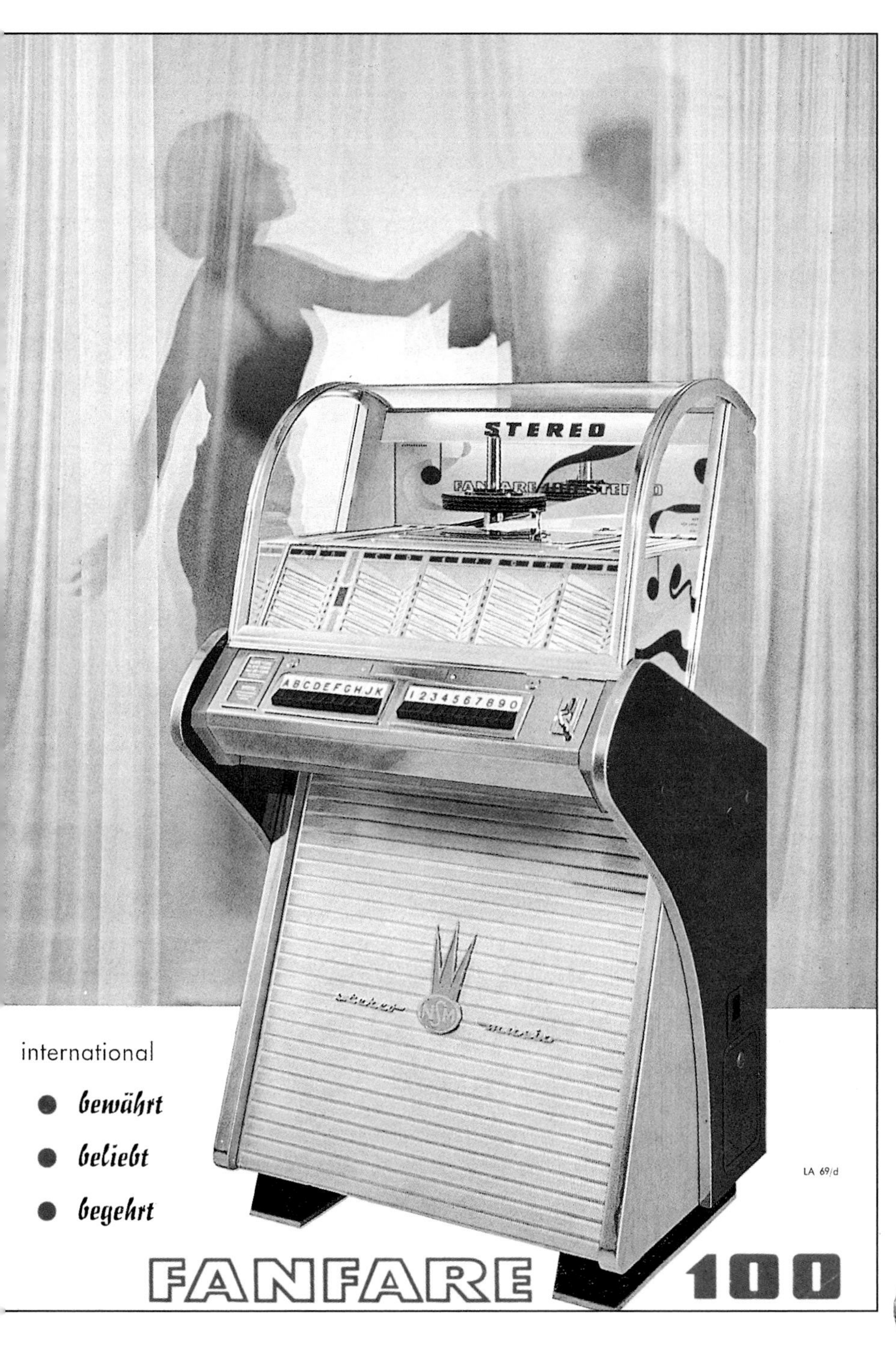

In appearance and in sound reproduction the German jukeboxes were equal to their American models.

clear: At that time, many letters from consumers reached us, saying nothing good about the Rotamint. It is really tempting to publish them today, for they would doubtless be seen as curiosities, now that it is clear that the Rotamint subsequently proved to be an economical and dependable musical device. In its way, though, it was as new then as the Fanfare is now, and everything new must begin with overwhelming prejudice and skepticism. It will only establish itself when the expert has become familiar with it and able to recognize, and also to evaluate, the advantages it offers."

With the "Fanfare 60" in 1956, the NSM firm introduced its first jukebox. The company, which had moved from Braunschweig to Bingen, was able to celebrate its fifth anniversary with 386 of its customers.

In the middle of 1958, NSM, in Bingen on the Rhine, introduced the new "Fanfare 100" model. In a press release, the NSM firm wrote: "We are very proud of our new Fanfare 100, which will undoubtedly continue the successful series begun by its predecessor, the 60, for all the experience gained from the construction of the noteworthy series of 5000 jukeboxes has been applied to the design and construction of the new box. Only after intensive tests had shown that the location of 50 records on the vertical storage column would cause no troubles, were the preparations for designing the 100 version finally undertaken."

The styling of the Fanfare 100 was reminiscent of the Seeburg V 200 of 1955, but was not as opulently ornamented with chrome and generally made a more spacious impression. This box was very modern for German conditions, and with

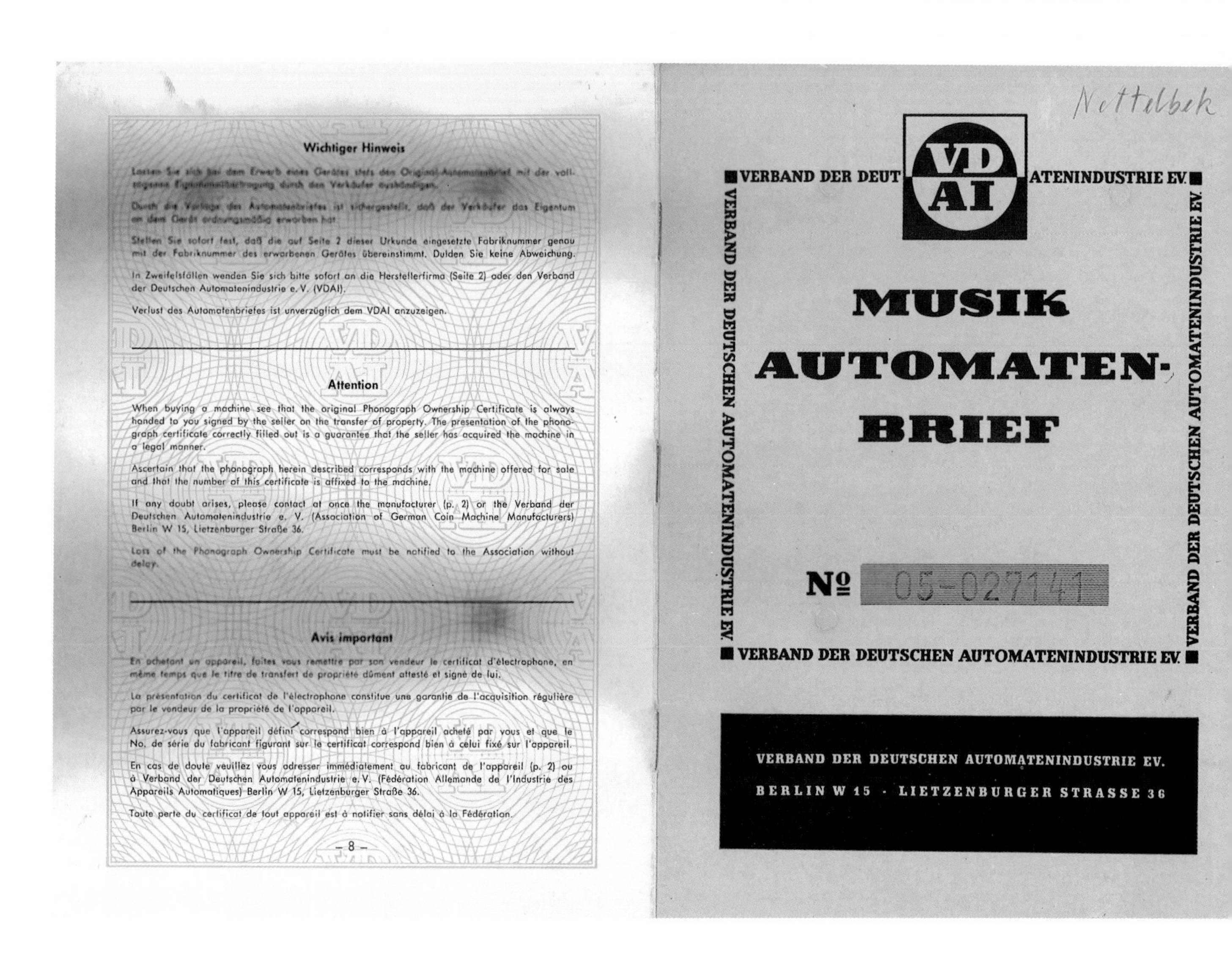

Wichtiger Hinweis

Lassen Sie sich bei dem Erwerb eines Gerätes stets den Original-Automatenbrief mit der vollzogenen Eigentumsübertragung durch den Verkäufer aushändigen.

Durch die Vorlage des Automatenbriefes ist sichergestellt, daß der Verkäufer das Eigentum an dem Gerät ordnungsmäßig erworben hat.

Stellen Sie sofort fest, daß die auf Seite 2 dieser Urkunde eingesetzte Fabriknummer genau mit der Fabriknummer des erworbenen Gerätes übereinstimmt. Dulden Sie keine Abweichung.

In Zweifelsfällen wenden Sie sich bitte sofort an die Herstellerfirma (Seite 2) oder den Verband der Deutschen Automatenindustrie e. V. (VDAI).

Verlust des Automatenbriefes ist unverzüglich dem VDAI anzuzeigen.

Attention

When buying a machine see that the original Phonograph Ownership Certificate is always handed to you signed by the seller on the transfer of property. The presentation of the phonograph certificate correctly filled out is a guarantee that the seller has acquired the machine in a legal manner.

Ascertain that the phonograph herein described corresponds with the machine offered for sale and that the number of this certificate is affixed to the machine.

If any doubt arises, please contact at once the manufacturer (p. 2) or the Verband der Deutschen Automatenindustrie e. V. (Association of German Coin Machine Manufacturers) Berlin W 15, Lietzenburger Straße 36.

Loss of the Phonograph Ownership Certificate must be notified to the Association without delay.

Avis important

En achetant un appareil, faites vous remettre par son vendeur le certificat d'électrophone, en même temps que le titre de transfert de propriété dûment attesté et signé de lui.

La présentation du certificat de l'électrophone constitue une garantie de l'acquisition régulière par le vendeur de la propriété de l'appareil.

Assurez-vous que l'appareil défini correspond bien à l'appareil acheté par vous et que le No. de série du fabricant figurant sur le certificat correspond bien à celui fixé sur l'appareil.

En cas de doute veuillez vous adresser immédiatement au fabricant de l'appareil (p. 2) ou à Verband der Deutschen Automatenindustrie e. V. (Fédération Allemande de l'Industrie des Appareils Automatiques) Berlin W 15, Lietzenburger Straße 36.

Toute perte du certificat de tout appareil est à notifier sans délai à la Fédération.

– 8 –

VERBAND DER DEUTSCHEN AUTOMATENINDUSTRIE EV.

VDAI

MUSIK AUTOMATEN-BRIEF

Nº 05-027141

VERBAND DER DEUTSCHEN AUTOMATENINDUSTRIE EV.
BERLIN W 15 · LIETZENBURGER STRASSE 36

A rare document in the languages of the western allies: a juke-box letter (probably from the early fifties) that allows the illustrated box to be used.

a different mechanism it probably would have found favor in America too. In 1960, thirty "Fanfares" a day were produced, and at that time they were exported to 27 countries.

The changing mechanism that was utilized in the Fanfare was similar to those used in record players for home use. There up to ten records were held on a spindle, and one after another dropped to be played. If the record player had not been used for some time, it was quite possible that two or three records could drop at the same time. The Fanfare mechanism, which had to accomplish considerably more than just letting records drop, caused similar problems.

The paying system is built on a purely electrical basis, with only relays doing the switching. The selection is given for all 100 possible choices. The selection indicator is equipped with a completely new effect. It is optically enlarged from a small band of film and projected onto a matte panel."

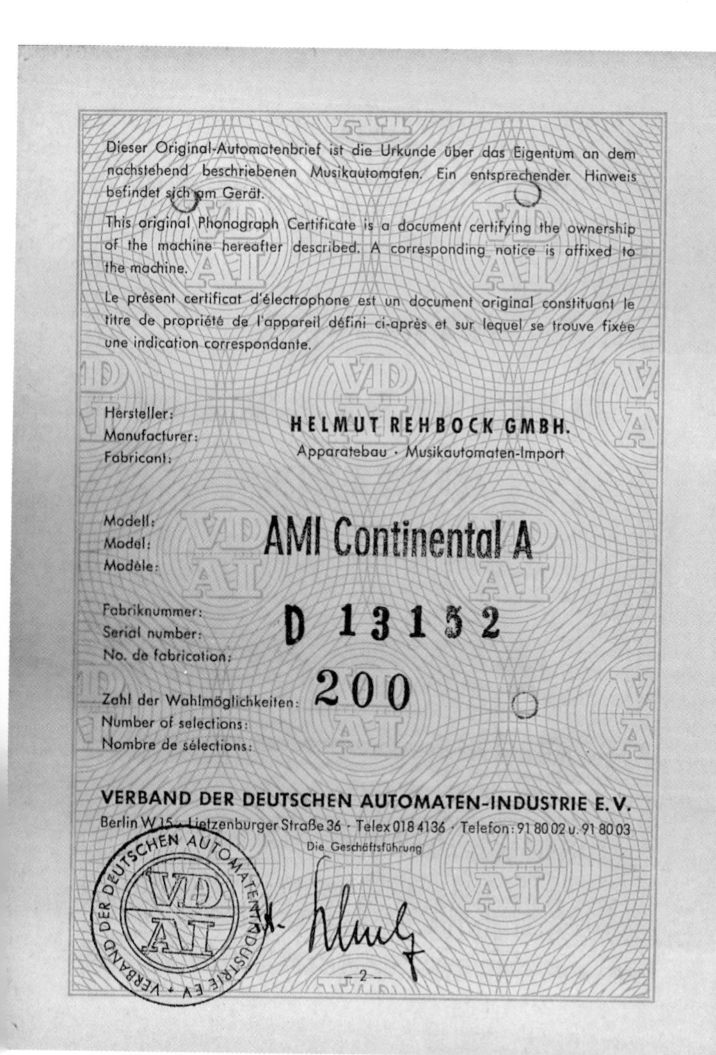

Dieser Original-Automatenbrief ist die Urkunde über das Eigentum an dem nachstehend beschriebenen Musikautomaten. Ein entsprechender Hinweis befindet sich am Gerät.

This original Phonograph Certificate is a document certifying the ownership of the machine hereafter described. A corresponding notice is affixed to the machine.

Le présent certificat d'électrophone est un document original constituant le titre de propriété de l'appareil défini ci-après et sur lequel se trouve fixée une indication correspondante.

Hersteller:
Manufacturer:
Fabricant: **HELMUT REHBOCK GMBH.**
Apparatebau · Musikautomaten-Import

Modell:
Model:
Modèle: **AMI Continental A**

Fabriknummer:
Serial number:
No. de fabrication: **D 13152**

Zahl der Wahlmöglichkeiten:
Number of selections:
Nombre de sélections: **200**

VERBAND DER DEUTSCHEN AUTOMATEN-INDUSTRIE E.V.
Berlin W 15 · Lietzenburger Straße 36 · Telex 0184136 · Telefon: 91 80 02 u. 91 80 03
Die Geschäftsführung

VERBAND DER DEUTSCHEN AUTOMATEN-INDUSTRIE E.V.
VDAI

– 2 –

№ 05-027141

– 3 –

Favorit

The Favoritbox was put on the market in 1957. The front of the box had flowing lines, the record carousel was very easy to see. The manufacturer of Favorit jukeboxes (in Hagen, Westphalia) gave out the following press release:

"Our new music box is in every respect a new design, which differs completely from the systems imported from America. To play the records, they are not taken out of a stack of records, rather the selected record in a carousel is simply turned in the direction of the player and played there. It is worth noting that no servicing or lubrication is necessary.

The Germans did not dream only of "Sole mio" and Capri: in 1955, Caterina Valente lured her listeners to Paris, which apparently dreamed only of love.

Lower left:
Federal Economic Minister Erhard and Governing Mayor of Berlin Willy Brandt listen to stereophonic records at the Philips stand during the Berlin Industrial Exposition of 1955.

Right page:
Mona Baptiste's smokey voice sang of Johnnie's birthday, and Horst Winter's dance orchestra reminded the Germans of the times before the war and the Carlton Bar in Berlin.

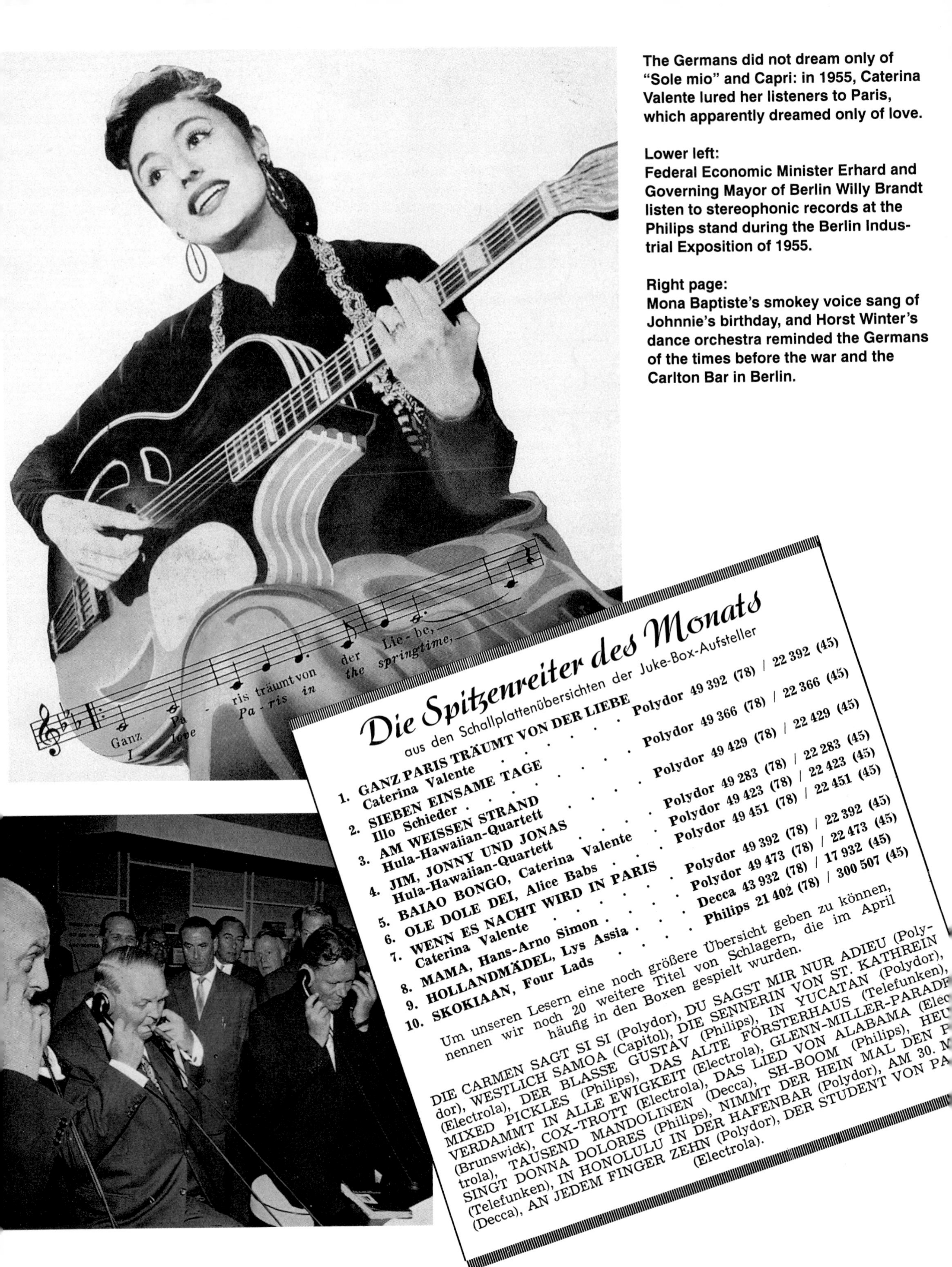
Ganz Pa - ris träumt von der Lie - be,
I love Pa - ris in the springtime,

Die Spitzenreiter des Monats

aus den Schallplattenübersichten der Juke-Box-Aufsteller

1. **GANZ PARIS TRÄUMT VON DER LIEBE**, Caterina Valente . . . . . Polydor 49 392 (78) / 22 392 (45)
2. **SIEBEN EINSAME TAGE**, Illo Schieder . . . . . Polydor 49 366 (78) / 22 366 (45)
3. **AM WEISSEN STRAND**, Hula-Hawaiian-Quartett . . . . . Polydor 49 429 (78) / 22 429 (45)
4. **JIM, JONNY UND JONAS**, Hula-Hawaiian-Quartett . . . . . Polydor 49 283 (78) / 22 283 (45)
5. **BAIAO BONGO**, Caterina Valente . . Polydor 49 423 (78) / 22 423 (45)
6. **OLE DOLE DEI**, Alice Babs . . . . . Polydor 49 451 (78) / 22 451 (45)
7. **WENN ES NACHT WIRD IN PARIS**, Caterina Valente . . . . . Polydor 49 392 (78) / 22 392 (45)
8. **MAMA**, Hans-Arno Simon . . . . . Polydor 49 473 (78) / 22 473 (45)
9. **HOLLANDMÄDEL**, Lys Assia . . . . . Decca 43 932 (78) / 17 932 (45)
10. **SKOKIAAN**, Four Lads . . . . . Philips 21 402 (78) / 300 507 (45)

Um unseren Lesern eine noch größere Übersicht geben zu können, nennen wir noch 20 weitere Titel von Schlagern, die im April häufig in den Boxen gespielt wurden.

DIE CARMEN SAGT SI SI (Polydor), DU SAGST MIR NUR ADIEU (Polydor), WESTLICH SAMOA (Capitol), DIE SENNERIN VON ST. KATHREIN (Electrola), DER BLASSE GUSTAV (Philips), IN YUCATAN (Polydor), MIXED PICKLES (Philips), DAS ALTE FÖRSTERHAUS (Telefunken), VERDAMMT IN ALLE EWIGKEIT (Electrola), GLENN-MILLER-PARADE (Brunswick), COX-TROTT (Electrola), DAS LIED VON ALABAMA (Electrola), TAUSEND MANDOLINEN (Decca), SH-BOOM (Philips), HEUTE SINGT DONNA DOLORES (Philips), NIMMT DER HEIN MAL DEN P… (Telefunken), IN HONOLULU IN DER HAFENBAR (Polydor), AM 30. M… (Decca), AN JEDEM FINGER ZEHN (Polydor), DER STUDENT VON PA… (Electrola).

The "Golden Fifties"

All the countries that took part in World War II experienced an unparalleled boom in the fifties. Germany too enjoyed the "Golden Fifties." The war was already over for some time; its horrors had been largely overcome and its damages had been rebuilt extensively.

The economy began to flourish; the people had work and money—and they wanted to spend it. This in turn helped the economy, which produced even more and could hire even more people. In short, the people had a better life for the first time in years.

All the people had plenty to eat, and most of them had jobs. In the stores, one could buy whatever one wanted. In one's pay envelope there was enough left over to spend on what one wanted.

To be sure, World War II had—especially in Germany—blown a hole at a very different level, and it was not so easy to fill. The old system of values had disappeared or been abolished, but there were no new ones. The "strong" Germany of Imperial days, with its overseas colonies, had not existed for many years and the "looking back" to German virtues that had followed the somewhat unstable Weimar era had led Germany directly into the abyss.

In the postwar years there was therefore a vacuum, particularly for young people. The older people, busy with the rebuilding of Germany, presumably found less lacking. The young people, on the other hand, had lost all their idols: Kaiser and Führer no longer meant anything.

Mona Baptiste

Horst Winter

There was nothing for them to believe in. The older people were despised for their self-satisfied attitude, and the concept of the "bourgeois" came back. The young people did not want to be like them. Or listen to their music—absolutely not!

That, short and simple, was the situation in the fifties, when the young people decided to seek their own idols. They found them in the music and the stars of the times.

But on the other hand, the defeated Germans also had something to offer the occupying forces. Pretty girls, for example—the "deutsche Fraulein" went over big with the GIs. And it was precisely these GIs who made the jukebox known in Germany. In their bars and taverns they played the music that the American occupying forces could not live without, no matter how far from home they were.

In the fifties, the jukebox embodied the outlook on life of an entire generation of young people. Their mobility was by no means as great as it is today. Thus even in the smallest villages there were "dance cafes," which were the young people's meeting places. Here they got together, here they could hear the music their parents decried. It was the high point of the "open" jukeboxes, which exposed the selection mechanism so the observer had something to see. Even before that, the exposed selection mechanism had been a treat for the eyes when it went into action, chose the record out of the stack and laid it on the turntable. Wurlitzer and Rock-Ola

Platten, die Geld bringen

78 er	45 er	
F 43 958	D 17 958	Mandolino / Simonetta
F 43 932	D 17 932	Holland-Mädel / Heute klopft mein Herz
F 43 828	D 17 828	Liebeswalzer / Tausend Mandolinen
A 11 571	U 45 571	Das alte Försterhaus / Kathrein
F 43 854	D 17 854	Holzhacker-Dixie / Warum fragst du immerzu, ob ich dir treu bin
F 43 944	D 17 944	Die Sennerin von St. Kathrein / Wenn ich die Berge seh'
F 43 946	D 17 946	Kleine Orangenverkäuferin / Mein schönstes Lied
F 43 637	D 17 637	Heideröslein / Friesenmädel
F 43 841	D 17 841	Heut' singen die Gitarren / Jim, Jonny und Jonas
F 43 754	D 17 754	Schweden-Mädel / Du bist Musik
F 43 768	D 17 768	Kleine Nachtigall / Der Kuckuck ruft
F 43 942	D 17 942	Schiffsjungentanz / Kleine Jodlerin
A 10 973	U 45 973	O mein Papa / Ponylied
F 43 940	D 17 940	Der Student von Paris / Sieben einsame Tage
A 11 628	U 45 628	Der alte Schäfer / Singe dein Lied, kleine Lerche
F 43 829	D 17 829	Cuculino / La Signora Musica
A 11 594	U 45 594	Wodka-Fox / Im grünen Wald
A 11 664	U 45 664	Im grünen Wald / Es wollt' ein Mädchen früh aufstehn
F 43 599	D 17 599	Du schwarzer Zigeuner / Melodia
A 11 671	U 45 671	Ganz Paris träumt von der Liebe / Ich werde von dir träumen
A 11 616	U 45 616	Schuster, bleib' bei deinen Leisten / Heut' liegt was in der Luft
A 11 672	U 45 672	Sphärenklänge / Dorfschwalben aus Österreich
C 2479	CF 2479	I Love Paris / Gigi
F 43 614	D 17 614	Jodeln kann ich nur, wenn ich verliebt bin / Tango der Nacht

Fordern Sie noch heute unseren kostenlosen Music-Box-Dienst an. Sie finden darin eine Zusammenstellung aller für Ihren Musik-Automaten bestgeeigneter Platten.

Teldec Schallplatten-Gesellschaft m. b. H., Hamburg 36

TELEFUNKEN
DECCA
CAPITOL
SCHALLPLATTEN

were aware of this and chrome-plated many parts for the sake of greater attraction.

Stars and Records

If one looks at the assortment of musical titles that should, from the standpoint of the supplier, be in a jukebox, one sees that the choice was not a simple one. Naturally, he could never know for sure who would visit the place or which records the people would want to hear. One sure choice was always the "hits," popular favorites that were in vogue at the time, and with which everybody was familiar.

In the trade journals of the jukebox distributors, their own hit parades, which listed the most popular titles, were published. And this shows that the public's possible choices from a jukebox were not limited to those 20, 48 or 100 titles that were found in the jukebox at any one time, but that a more extensive choice was made. The press on the selection button of the jukebox was simultaneously the choice for or against certain pieces of music. The supplier had to proceed according to the public's wishes—and they could be very different from one locality to another. What great significance was attributed to the public's wishes became clear in a contemporary article in the "Automatenmarkt": "The public wants to hear *its* melodies, but it usually links the melodies with *particular* names, voices or vocal styles. The supplier absolutely must consider this situation when choosing records. The demands of the public are, in this case, *categorical.* Experience has taught that it simply refuses other recordings."

Left:
The records that TELEFUNKEN recommended for jukeboxes.

For the supplier, then, it was vitally important always to supply the right records; one spoke at that time of the correct "programming" of jukeboxes. Hit parades were therefore important, not just to the public. Hit lists for jukeboxes, such as were to be found in "Automatenmarkt," were meant to provide the suppliers of jukeboxes with helpful information as to titles that were especially popular. If one takes a good look at hit lists from that era, it becomes clear that big names dominate the scene. The silent majority also used the jukebox, and it was not just Elvis Presley and Bill Haley that were wanted and played. A list from August 1954 looked like this:

1. *Das Heideröslein*, Friedel Hensch, POLYDOR
2. *Das Schwedenmädel*, Lys Assia, DECCA
3. *Das alte Försterhaus*, Rodgers Duo, TELEFUNKEN
4. *Bella Bimba*, Bibi Johns, ELEKTROLA
5. *Diesmal muss es Liebe sein*, G. Griffel, TELEFUNKEN
6. *Es liegt was in der Luft*, Mona Baptiste, POLYDOR
7. *Oh, mein Oapa,* Lys Assia, DECCA
8. *Der Autoreifen*, J. Peheiro, DECCA
9. *O Cangaceiro*, H. Zacharias, POLYDOR
10. *Bon soir, bon soir*, Vico Torriani, DECCA

The hit parade still was called the "Schlagerparade" in Germany at that time. According to an interesting story, there were times when the jukeboxes were so widespread that one could scarcely avoid them—or the music they made. That was just the right time for a record called "Silent," a record without any sound, that promised three minutes of healthy rest from the noise of the jukebox. In the trade paper "Automatenmarkt," from which the hit parade printed above comes, there was an advertisement a few pages before, offering this Polydor record: "Drei Minuten Pause" (Three-Minute Pause). In the USA, there was not a jukebox without this record. "Issued exclusively for jukeboxes."

The hit songs of those days were sung by performers whose voices are, in part, still known to us today: Helmut Zacharias, Caterina Valente, Bibi Johns, Peter Alexander, Gerd Wendland, Giesla Griffel, Lieslotte Malkowsky, Les Paul, Mona Baptiste, Peter Kraus, Anette, Petula Clark, Ruth Berle, Dany Mann, Andrey Arno, Bert Kämpfert, Hazy Osterwald Sextett, Fred Rauch, Freddy Quinn and so on.

Many of these stars utilized their popularity in films too, mostly in comedies. Even if the hits of those days seem sort of old-fashioned today, they were very successful then. For example, "Oh my Papa," performed by Ray Anthony, was number one in America for weeks.

Again and again there were also hits that referred directly to the jukebox. "Hast du nicht 'nen Groschen für die Musikbox" and "Die Musikbox" are examples of this, as is the juke-box hit "Für zwei Groschen Musik," with which the performer Margot Gielscher won the first prize in the German "Grand prix d'Eurovision 1958" competition. In a report in "Automatenmarkt," the importance of the jukebox at that time is made very clear:

> The text of the refrain expresses so accurately the real spirit of the jukebox without which one cannot imagine our present-day life. In a few lines, it is expressed here that the jukebox offers many people an atmosphere of relaxation and a little distraction every day at low cost, and that its melodies bring happiness and cour-

age to countless people every day in an often unfriendly everyday world:

Für zwei Groschen Musik *Music for two dimes*
Für zwei Groschen musik *Music for two dimes*
Und für so wenig Geld *And for so little money*
Gehört euch eine Welt *A world belongs to you*
Für zwei Groschen Musik *Music for two dimes*
Für zwei Groschen Musik *Music for two dimes*
Und der Alltag versinkt *And every day disappears*
Wenn froh Musik erklingt *When happy music sounds*

Thus the quotation from the "Automatenmarkt." Three years later (1957), the juke-box suppliers' hit parade looked like this:

1. *Cindy, oh Cindy*, Margot Eskens, POLYDOR
2. *Ich weiss, was dir fehlt*, Peter Alexander, POLYDOR
3. *Was kann schöner sein*, Lys Assia, DECCA
4. *Sei zufrieden mit dem Heute*, Rodgers-Duo, ODEON
5. *Weisser Holunder*, Gitta Lind, DECCA
6. *Just walking in the rain*, Johnnie Ray, PHILIPS
7. *Peter, komm heut abend zum Hafen*, Die Sunnies und die Correlis, TELEFUNKEN
8. *Singing the Blues*, Guy Mitchell, PHILIPS
9. *Mauerblümchen*, Erni Bieler, POLYDOR
10. *Que sera, sera*, Doris Day, PHILIPS

Even if the assortment is certainly not representative, still an obvious "Americanization" of musical taste can be seen.

At the beginning of the fifties, most of the hits that were played in German jukeboxes were still produced in Germany, including "Germanized" stars such as Peter Alexander (Austria) and Vico Torriani (Italy), and others who almost always sang in German, though with an (enchanting) accent. Even the

following twenty most-played records listed there were overwhelmingly German.

But this changed with the years. By 1957, American titles were already found among the top ten, and the next twenty also included titles like "Tutti-Frutti" and "See you later, Alligator." Performers such as Elvis Presley, Chuck Berry and Bill Haley—and many others—defined the direction of modern music. Rock 'n Roll had arrived.

Below:

Even though there was not yet a concept of "subculture," looking at such photos seems to inspire the impression that there was something like a culture of its own around the jukebox...

The right choice of records by the supplier was made much easier by a device in the jukeboxes: Many of the new boxes at that time—the mid-fifties—were equipped with a so-called popularity gauge. With it, a popularity curve was set up for every single record, so that the supplier could see exactly how the hits and the flops looked in this particular jukebox, even if it did not show which hits one or another guest would have liked to hear more often. Unpopular titles, on the other hand, could be replaced, so as to try to inspire the people to play more often and thus spend more money.

In any case, it was not as easy to earn money with jukeboxes as it may have seemed. At first the boxes were fed with ten-pfennig pieces. With the jukebox selling new for 6000 marks in Germany, that meant that the buyer needed 60,000 (!) playings just to pay for the box. And this did not take the running costs (supplying with records, repairs, etc.) into consideration. The buyers in the USA had it much easier, as they only needed to pay about a thousand dollars for a similar jukebox. The guest had to put five cents into it, so it took "only" 20,000 playings to amortize a jukebox. The German owners very soon began to charge 20 pfennig, to have the jukebox at least paid for after 30,000 playings.

This calculation also shows how popular the jukebox must have been, for after all it was a business

Spielhölle in Stadtmitte

Zehn Groschen machen eine Mark

„Spielhallen" verführen die Jugend — Treff...

...ungslose Eltern / Von unserem ständigen Mitarbeiter

das Jugendamt bestätigen die Gerichtsberichte: Die Spielhallen sind häufig Ausgangspunkt für Ueberfälle und Raubzüge in den Straßen der nächtlichen Großstadt. Da werden keine großen Dinger gedreht, von Ausnahmen abgesehen: Zigarettenbuden stehen auf dem Programm, keine Juwelierläden. Man braucht ja nur Groschen.

So forsch sie nach außen auftreten, die labilen Naturen überwiegen unter den Spielhallenbesuchern. Sie sind „autoritätsgläubig". Wehe, wenn jemand an ihrer „Männlichkeit" zu zweifeln wagt! „Du traust dir ja nicht" — mit dieser verächtlich hingeworfenen versteckten Aufforderung beginnt hier allzuoft eine Laufbahn, die hinter schwedischen Gardinen endet. Und die Eltern haben keine Ahnung von der Gesellschaft, in die ihr Kind „spielend" geraten ist. Es ist kein Zufall, daß sich die Spielhallen durchweg in solchen Vierteln der westdeutschen Großstädten befinden, die man nach Einbruch der Dunkelheit nur widerwillig betritt ...

Das neue Jugendschutzgesetz, das im Januar 1952 in Kraft trat, beschäftigt sich in einer einzigen Bestimmung mit den öffentlichen Spielhallen. Nach Paragraph 7 darf Jugendlichen unter 16 Jahren der Zutritt nicht gestattet werden. Vor zwei Jahren waren Spielhallen so gut wie unbekannt. Heute täte es dringend not, das Alter auf 18 Jahre heraufzusetzen!

Gefährlicher Zauber rotierender Zahlen

Spielhallen als „Hintertür" des Gesetzes / Künftig nur noch drei Automaten mit Gewinnchance?

Jugendgefährdende Spielhallen

Spielwut gefährdet Jugend

VON LISELOTTE WEBER

Bonner Jugend will keine „Spielhalle"

Hagen: Kein „Monte Carlo des kleinen Mannes"

Einzige Spielhölle in Hagen wurde vor einiger Zeit geschlossen

Das Dorado der Halbwüchsigen

Frankfurt hat genug Spielhallen / Gefährdete Jugend

Der große Gewinn wird in den Spielhallen gemacht. Und so gering die Vorwürfe gegen das Spiel an dem einem oder den zwei Automaten in den Wirtshäusern sind, so massiv führt man sie an und so leidenschaftlich wird zurückgekontert, wenn es um die Spielhallen geht, in denen die Automaten zu Dutzenden hängen.

Spielbank des kleinen Mannes

„Groschenspiele"

Das ist der Anfang! Haltet die Kinder hier weg.

Eine Sorge der Mütter

Spielleidenschaft ein...

Ist der Kleine Lunapark ...

Spielhöllen gegen...

Razzia auf St. Pauli

Nervenkitzel am Spielautomaten

Jugend in Gefahr

Gemischte Gesellschaft im Spielsalon - Sechzehnjährige vor dem ...

Verordnung der Bundesregierung in Vorbereitung:

Spielhallen sollen geschlossen werden

Wann tritt das neue Gesetz in Kraft? — „Stützen des Staates" mit großem Verdienst

In Lübeck und in Travemünde gibt es insgesamt vier Spielhallen. An der Frage der Notwendigkeit dieser Betriebe erregen sich die Gemüter in der Hansestadt. Verschiedener Meinung in dieser Frage sind auch so hohe Institutionen des Staates, wie das Bundesfinanzministerium und das Bundesinnenministerium, von denen das erstere in den Spielhallen eine willkommene Geldquelle erblickt, während das letztere der Ansicht ist, man müsse unbedingt von Staats wegen derartige Einrichtungen verbieten, da diese eine Gefahr für die öffentliche Moral und die öffentliche Ordnung bildeten. Noch aber liegen die Ministerien miteinander in dieser Frage im ...

schon dafür, daß keine Jugendlichen unter 16 Jahren in den Spielsälen aufkreuzten und daß keine jungen Leute unter 18 mit ihrem Groschen die Automaten in Bewegung setzten, denn wie das Damoklesschwert hänge über den Spielhallen stets der Satz von der Gefährdung der öffentlichen Ordnung. Schließlich hätte man auch vom Ordnungsamt aus stets ein sehr wachsames Auge auf die Spielsäle, unterstützt von

that would only continue as long as it made a profit. And a jukebox that required "only" 30,000 playings was still a business loss. On the average, a jukebox would probably have played about 100,000 records by the time it was retired.

In those days, the jukeboxes were usually made available to the host without cost, and of course he did not receive any percentage of the profit. It was a kind of reciprocal undertaking in which the owner invested a lot of money in a jukebox and wanted to be as sure as possible of getting his money back. Sharing the profits was simply not profitable for him. On the other hand, he depended on the owner of the place, for only there could he set up his jukebox, only there could he reach the people who were willing to spend money on it. But the host also gained something from the box, not money, but a box that provided music. And that in turn drew the people.

...which immediately inspired protective measures for youth and against the ranks of jukeboxes in the so-called "halls of play."

The Peak and Decline of an Era

Only for about ten years—from the early fifties to the early sixties—did the jukebox play a significant role in people's lives. This era is in the past now, for the radio as well as the jukebox. Both of them still exist, people still use them, but the important events take place elsewhere. Today the latest news does not come over the radio any more, but on television.

But that is another matter. The jukebox had reached its high point in the mid-fifties. In America there was a jukebox for approximately every 330 Americans. The names of AMI, Seeburg, Rock-Ola and Wurlitzer were known to all, and the 8000 users in the United States bought some 74 million records per year. In the next two decades, the ratio of jukeboxes to the public sank to one in 530. Thus the bastion of juke-box production was certainly to be found in America. Only Germany was also able to produce jukeboxes in numbers worth mentioning. Although Wurlitzer and Rock-Ola opened factories in Europe, German jukeboxes were able to gain a very significant part of the European market, and were even exported all over the world.

At first, though, this took place slowly. The year of 1950 was not yet the year of the jukebox in Germany. At that time there were exactly 22 jukeboxes, with a total value of $2697, imported from the USA (stated in "Automatenmarkt," 11/51). These were almost exclusively used machines produced in America before the war.

Germany's entry into the jukebox business was a slow one, for the buyers were at first reluctant to buy, as they feared a backlash in conservative regions and did not want to take on a large investment in jukeboxes without security. But by 1954 the picture had changed.

Related devices are found along with jukeboxes: chewing-gum and peanut machines, slot machines, neon signs and other advertisements. The one-armed bandit holds its hand out over them.
The laboriously assembled objects quickly give one the impression that one is looking at witnesses of a better time.
Large expositions—like this one at Kaunitz in 1993—allow the Germans to compare a really extensive number of such interesting devices in one place for the first time.

According to estimates, at the end of 1953 there were barely 1000 jukeboxes in operation in the Federal Republic of Germany. And this number increased dramatically: One year later, there were already 4500. Another twelve months later, the total had already increased to around 12,500 jukeboxes. By the middle of 1959, the total had tripled once again: about 40,000 jukeboxes were to be found in West Germany.

The German juke-box industry developed quickly and splendidly. In 1956, jukeboxes worth 2.3 million marks were already being exported; these were not only all-German products but also American products built under license in Germany. And in 1957, according to official statistics, 2701 jukeboxes, with a combined value of 6,045,006 marks, were exported from West Germany and West Berlin. In the same period of time, 25,779 jukeboxes with a total value of 16.4 million dollars were exported from America; to be sure, the greater part of the American exports consisted of used machines, very much in contrast to the German exports.

An interesting curiosity is the fact that East Germany did not close its borders to jukeboxes. As can be found in a 1959 report, there were then about 100 jukeboxes in HO restaurants there. Although the state trade organization of East Germany had to pay a high price for the boxes (10,000 eastern marks), it was probably a good deal; according to reports, there were restaurants that could gain up to a thousand eastern marks a week from such a box. As for what records these jukeboxes were filled with, that is another matter.

In the early sixties, the juke-box industry went downhill fast in the USA. Interestingly, the author of the English book "Jukebox Saturday Night," John Krivine, relates the decline of the jukebox to developments like McDonald's. In fast-consumption places, particularly fast food, he sees a reason why jukeboxes are no longer so interesting today. The people who formerly took their time over a meal, spending about an hour at it, also had the time to drop a nickel in the slot. There is no time to do that at Burger King, etc!

And he may not be so very wrong. The basic business principle that made jukeboxes so successful for the owner, namely that of bringing in as much as possible per hour, ironically contributed to their downfall in the end. Fast-food chains like McDonald's and Burger King also depend on a fast turnover, and not on people feeling good and staying there a long time.

Still in all, it would be too simple to push the blame for all of this onto fast food places alone. Krivine properly refers to other reasons for the decline of the juke-box market as well.

The greatest opponent was surely television. It keeps the people at home, provides free entertainment there, so that fewer and fewer people see any need to find pleasure in public houses.

With social changes taking place, there was a lack at all levels of suitable places to set up jukeboxes. The onetime dance cafes became discotheques, and in many restaurants the jukebox has given way to the speakers of the stereo set.

Stereophonic sound, developed and introduced at the end of the fifties, was also a factor in the decline of the jukebox, for it was not made or intended for stereo. The relatively compact jukebox could not provide music with an illusion of space, and if the speaker had been taken out of the jukebox and placed somewhere else in the room it would not have been a jukebox any more, although there were certainly attempts to do just that. The desire for greater realism in music and the resulting demands for higher quality could not, in the end, be met by the jukebox. The times called for something different. The charm of the music played by an "old" jukebox is something we are just rediscovering today.

The Design of Jukeboxes

The appearance of a jukebox was decisive for its success, for there were few technical advantages from one box to another. At first there were very few technical changes. The heart of a jukebox, understandably, is the record changer. And precisely this part was changed the least. The manufacturers used their changing mechanisms for years and decades, as long as they proved to be reliable. Wurlitzer produced nearly sixty models that utilized the "Simplex" mechanism. The AMI firm also used its selection mechanism for nearly thirty years—from 1927 until the early fifties. Technical arguments for or against one or another mechanism rarely surfaced. It merely had to grasp the right record, and as long as it did that reliably, there was no reason to change anything about it. Thus the appearance of a jukebox took on decisive importance, and for this reason the outsides of the boxes were changed constantly; the biggest firms put a new model on the market almost every year.

In terms of style, appearance and development of jukeboxes, three major periods can be recognized, the first being from the thirties to the beginning of the fifties (the Golden Age). In these two decades, music boxes became big. The following decade, the time from the early fifties to the beginning of the sixties (the Platinum or Silver Age), brought the high point of the jukebox. And then, in the years after that, came the decline of the jukebox.

The jukeboxes that were built before World War II all had a more or less fancy wooden case. Lavish decorations, trim and inlays, and polished surfaces, dominated the picture. In the forties, and especially in the fifties, lighting, translucent colored plastic panels and light effects dominated the scene, and in the fifties there was a very particular emphasis on an open, visible changing mechanism. This was undoubtedly the era of the jukebox, and there are even some people who see an essential relationship between the covering of the changing mechanism and the decline of the jukebox. This is surely a considerable exaggeration. But as the spirit of the times changed, the jukebox also changed—and it no longer went with the times. In the sixties it became increasingly more angular, with a clear basic shape—aluminum grilles and new "objectivity" dominated the picture.

In the course of these great style changes, there were often updating kits available, with which used jukeboxes could be modernized to suit the tastes of the times. For example, Seeburg offered the "universal cabinet," into which older jukeboxes could be fitted. Other such kits required that parts of the old box be removed or broken loose and replaced by new, modern parts. And there were the "add-ons," with which old jukeboxes could be enriched with light shows. And similarly to the way the style of jukeboxes changed, these modernizing kits also changed. In the forties and fifties, adapting kits were offered that livened up the outside of the box, adding elements—especially lights. But with the sixties there came a different direction. Now it was a process of removing unnecessary decorations from the box and giving it as smooth an appearance as possible.

Left:
Wurlitzer Model 35, made in 1936, with a well-built hardwood case.

Design Inspirations

The design of a jukebox took on an important task: It had to lure the guest to approach it, and more than that, it had to entice him to insert money—and not just once but again and again.

This in turn meant that on the one hand it had to suit the spirit of the times, and on the other, it had to be attractive enough to draw the eye toward it. Very early on, the manufacturers of jukeboxes recognized the possibilities offered by plastic and used it very willingly.

But in the fifties, jukeboxes were already in a very different phase of development. Their inspiration came from Detroit, more precisely, from the auto industry located there. If one looks at the jukeboxes of the fifties, especially those of the mid- and late fifties, and observes their shapes, then it can be seen immediately that the essential elements of automobile design have found a home in the jukebox. The "cruiser" found its counterpart in the jukebox: tailfins, back-up lights, radiator grilles, fenders, windshields—even foot pedals were regarded as sources of inspiration by the jukebox manufacturers and appeared in different forms in the jukeboxes.

Chrome dominated the picture. Defining elements of shape for the auto industry—and very soon also for the jukeboxes—came from the beginnings of space-travel technology. Inspired by the sleek styling of jet fighter planes, Cadillac introduced the first of the famous "tailfins" in 1955.

With these elements, enhanced by the stylish colors of the times (turquoise, light green, light blue, light pink, and a vivid red in combination with beige), the manufacturers of jukeboxes also decorated their models. This makes a very nice appearance on the Seeburg HF 100 G, all the decorative parts of which are chromed.

In 1954, Cadillac introduced the "panorama" windshield, which was to become another characteristic of the cruisers—and the jukeboxes. This was a wraparound windshield, curved at both sides to afford a better view. A year later, the rear windows also took on this shape.

In 1956 the Wurlitzer firm introduced its 1900 and 2100 models, the first jukeboxes to use this style element. In the mid-fifties the AMI jukeboxes were also strongly influenced by the automobile styling of the time—the wraparound windshield in particular was utilized. The panoramic glass used by AMI in 1957 was even more impressive than that used by Wurlitzer. And chrome plating was utilized more lavishly on this box than on any other.

Another style element taken from automobile design turns up on jukeboxes again and again: the V. In 1946, Cadillac became the first automotive firm to use the V emblem. Later it was also adopted by twelve other auto manufacturers in the USA, and in the mid-fifties it was very widespread. Rock-Ola first used it on the Model 1457 in 1957. On the otherwise very plain speaker grille the firm's own creation of a coat of arms was mounted, and below it, large and eye-catching, the V was placed. It symbolized speed and dynamics, and also stood as a symbol of the V-8 motors used in almost all American cars. Rock-Ola made the emblem out of plastic, not of chrome-plated metal, as it would not have stayed on the box for long otherwise; interested "collectors" would have made sure it disappeared in no time.

Seeburg took a very different route, at least in terms of external case form. In a day when everything had to be rounded, spirited and expansive, Seeburg introduced the KD series, jukeboxes that were kept very smooth and strict. But they too did not disregard elements from the auto or space industry. On the speaker grille, three chrome fins with red lights were mounted; they could have symbolized either the tailfins of a car or the control surfaces of a rocket.

With its Tempo, introduced in 1959, the Rock-Ola firm took up another element of automobile styling: the tailfin. They were eye-catching, these "tailfins" of chromed zinc alloy, mounted on the sides under the dome. Although Rock-Ola, with its Tempo and Tempo II, opposed the trend of the times toward simpler, cleaner lines (only AMI still built similarly lively boxes), the boxes were still very much appreciated—especially by the younger populace.

The Wurlitzer Model 700 of 1940 still had a comparatively flat "facade" with decor reminiscent of glass inlays or garden gates. The Model 1100 of 1948, on the other hand, begins to resemble the radiator grille of a Cadillac (here the "Coupé de ville") of about the same era.

The influence of space-travel technology is really obvious in the AMI "Continental"; its nose cone and radar screen for the titles are very reminiscent of the first successfully launched satellite, the Sputnik.

Along with the technical innovations of those times that flowed into the styling of automobiles and jukeboxes, the events taking place in the royal houses of the world influenced their contemporaries very strongly.

So it is surely no surprise that the crown emblem turns up again and again. In the most varied forms, it can be found on products from refrigerators to automobile emblems. Product names such as "Regina," "Princess" or "Empress" were also very popular among juke-box manufacturers; for example, there was the Rock-Ola "Regis."

And on the speaker grille of the German "Fanfare" jukebox too, a crown can clearly be seen. The Seeburg "KS 200" likewise shows stylized crown elements on its moving parts and side panels. These are only two examples of many uses of the crown symbol in the juke-box industry. The situation in the auto industry was very similar, as well as for many other devices used in daily life at that time.

The great interest in royalty in those times probably cannot be explained precisely any more. But there are several indications that make it clearer why royalty was so popular in those days:

- In 1952 Elizabeth II was crowned Queen of England.
- In 1956 Prince Rainier III of Monaco married the movie actress Grace Kelly. This event in particular attracted great attention at the time, and many have not forgotten that wedding to this day.
- Soraya, the wife of the Shah of Iran, was constantly in the limelight of press reporting in the mid-fifties.

The panoramic windshield of the American car was also found in the jukeboxes of the fifties. This is the Wurlitzer Model 2000 of 1956.

Wurlitzer
200 SELECTIONS
200

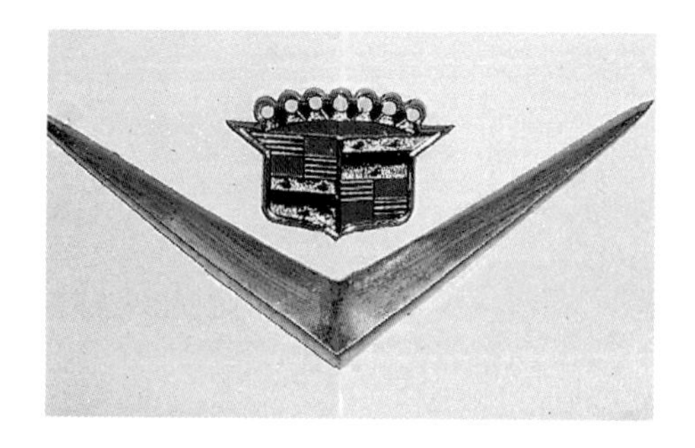

Door trim in the form of a projectile on a Ford Thunderbird of the mid-fifties, and the same projectile combined with the broad Cadillac emblem decorating this Seeburg jukebox.

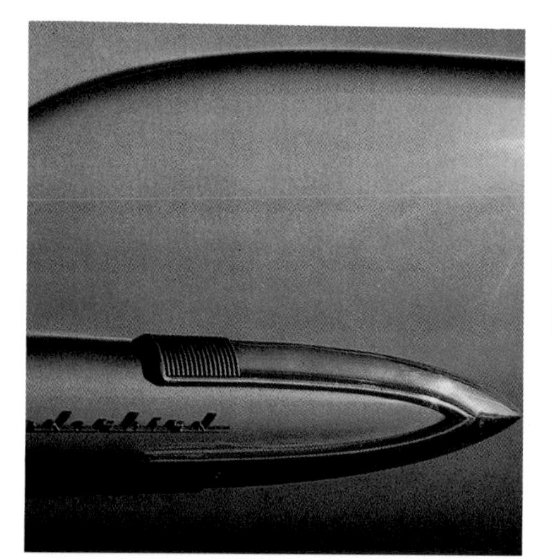

Compare these details of an American car of the fifties with details of the jukeboxes (starting on page 64)—the relationship between automotive and juke-box design is unmistakable. Both represent a "chromed" dream world. On the other hand . . .

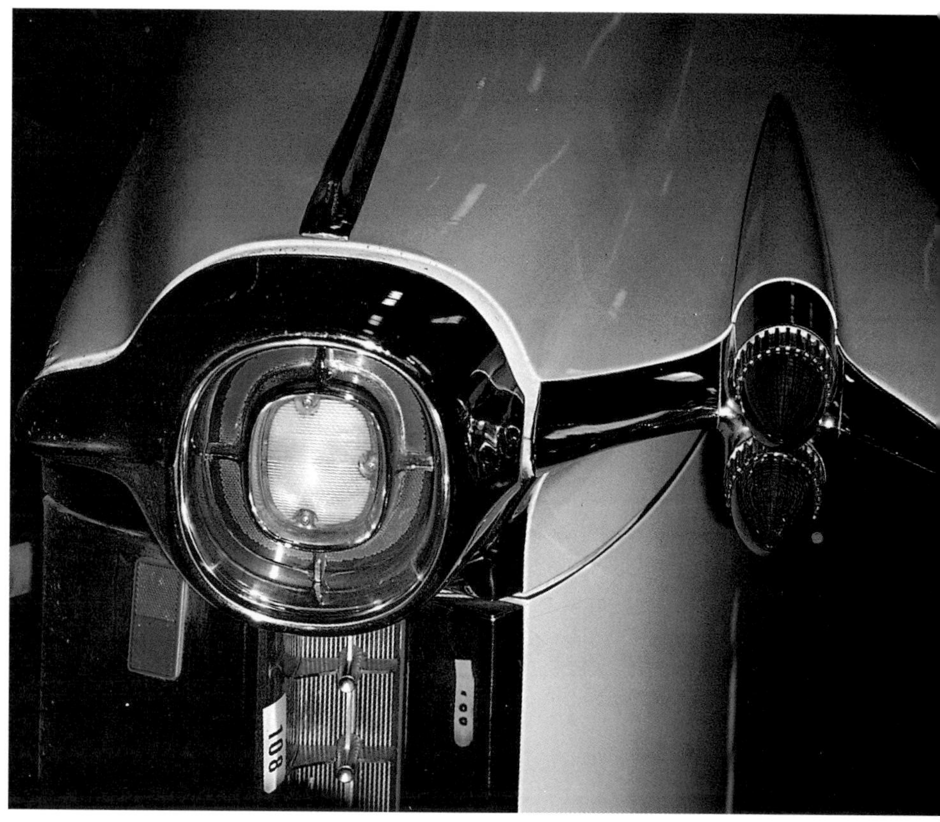

...the design of the "German living room" with kidney-shaped table and floor lamp, early asymmetrical shapes, and furniture of various kinds of wood, looks rather modest.

These are only a few outstanding events of those times. Along with them, there was naturally also the usual everyday gossip about the royal houses; in those days, as now, people were greatly interested. Royalty in any form thus moved the public. So it is presumably no surprise that these events were adapted by the designers of appliances.

At the beginning of the sixties, the automobile industry headed in a different direction: smooth surfaces, straight lines, and fenders extending far to the front or back dominated the appearance of the "side view" cars.

And the design of the jukebox also became increasingly more restrained and less eye-catching in the sixties. It no longer played any major role in young people's lives. Since the target group had become older, the manufacturers made efforts to build less intrusive jukeboxes.

With this unobtrusiveness, though, the necessity for the buyers to trade models in for more modern ones at regular intervals disappeared. These unobtrusive boxes could stay where they were much longer. And so they did, and the business turnover of the juke-box manufacturers decreased even more.

Page 62:
What is already a collectors' item as a piece of history stood in the middle of certain social rituals forty years ago. What the picture of the Wurlitzer 1450 clearly shows is that flirting could be done deliberately over the jukebox. In the process, it was important to choose the right music.

Model 1450

Gallery of Jukeboxes

Rock-Ola 1448

Rock-Ola 1448
Built in 1955
Made by: Rock-Ola Manufacturing Corporation, Chicago
Model 1448
Serial number 145 346
120 selections

The break with the traditional is complete. The new 120-selection box is glittering with chrome from top to bottom.
A triangular roller holds the title labels. When the red program button is pushed, the title holder rotates on its axis. The choice is made by pushing one of the forty buttons. Then the vertically mounted turntable is set in motion.
A pushing device prevents the records from falling out at the bottom. The gripper takes the desired record out of the magazine and places it on the turntable.
The chromed teeth on the speaker grille may have been taken from the Seeburg Model R.

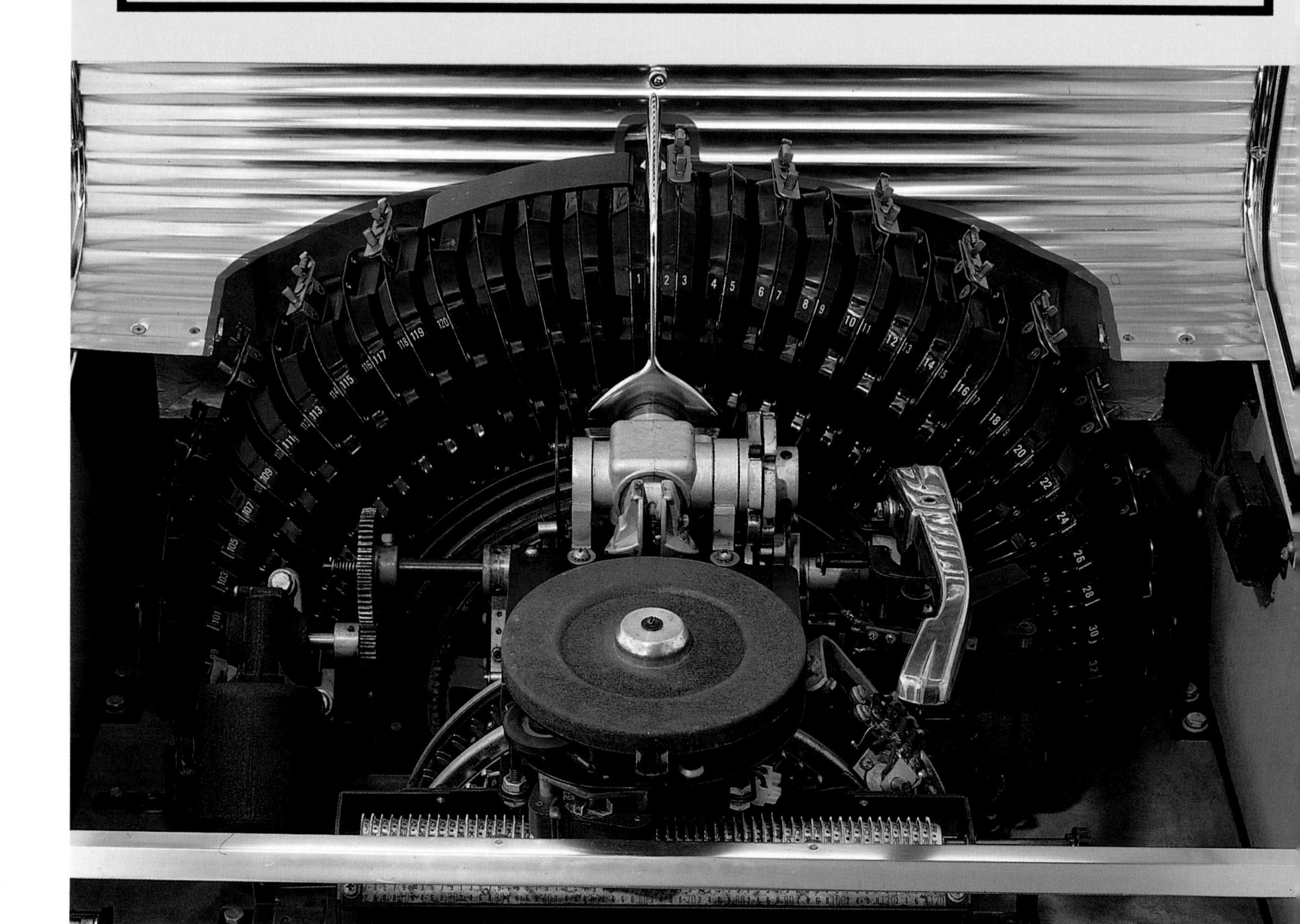

ROCK-OLA
AUSWAHL
PROGRAMM WECHSEL
HIGH-FIDELITY 120 SELECTIONS

Rock-Ola 1448

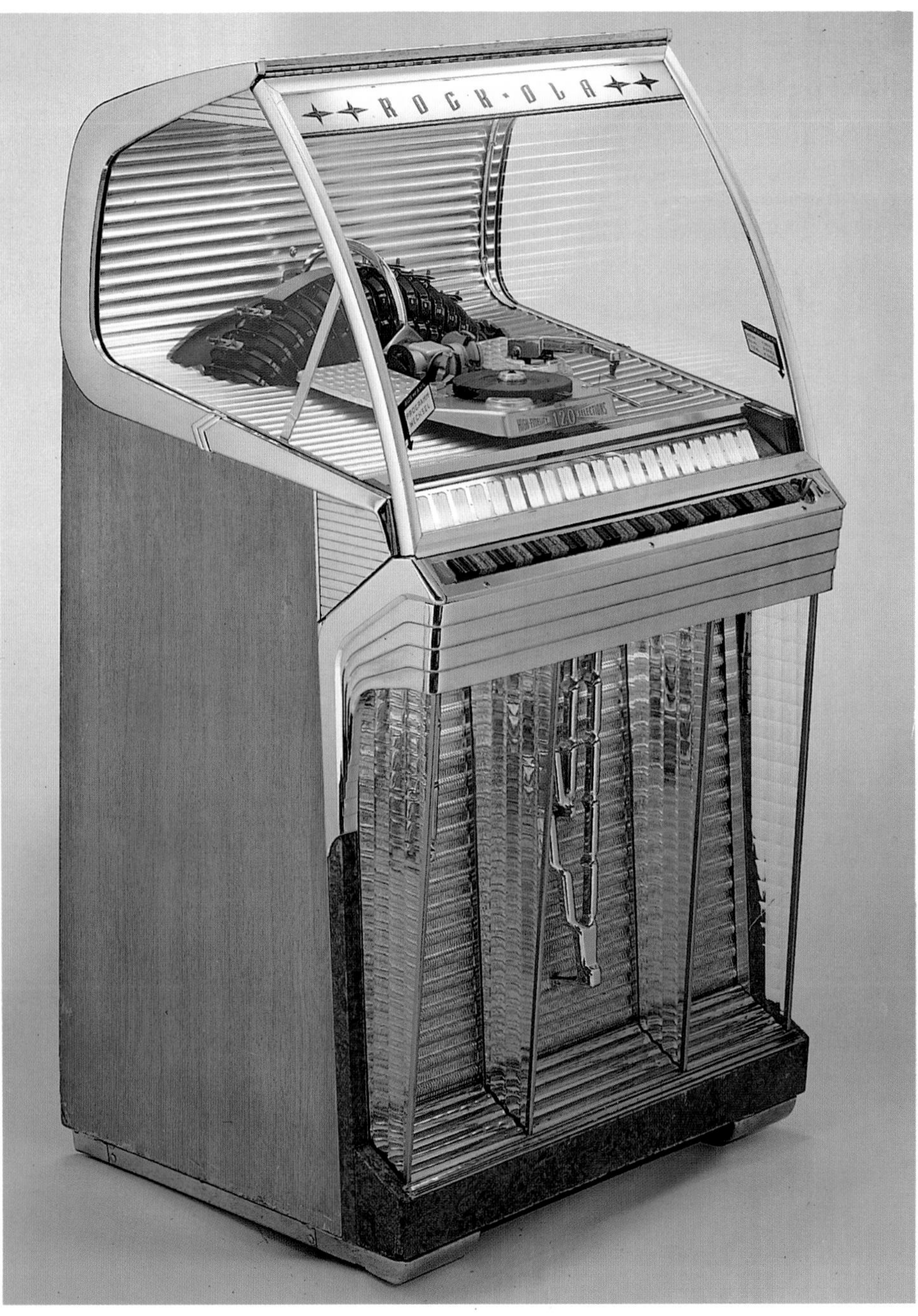

Seeburg R

Seeburg R
Built in 1954
Made by: J. P. Seeburg Corporation, Chicago, Illinois
Model HF 100 R
Serial number 136 321 3
100 selections

Six aggressively flashing teeth on the speaker grille make this chromed monster look downright dangerous. The machine with its blue top panel is set off majestically from the rippled gold background.

The two-tone wooden case, seen from the side, gives a lively impression. From this perspective too, the playing mechanism is easy to see. The choice of records is made by using ten letter and ten number keys.

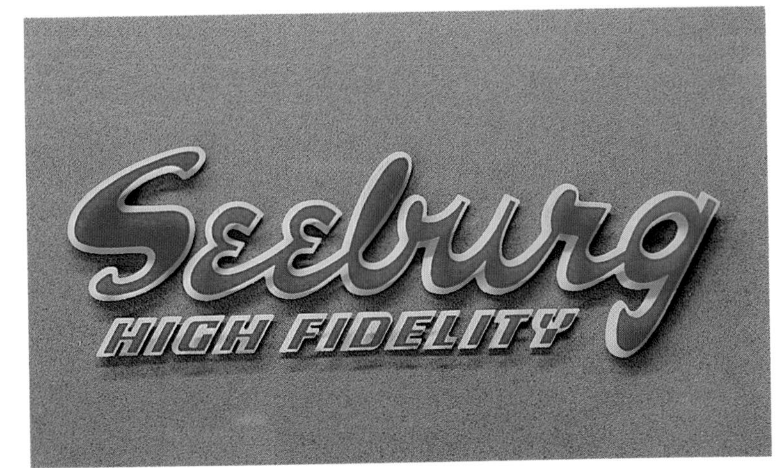
Seeburg
HIGH FIDELITY

100

D1 D3 D5 D7 D9 E1 E3 E5 E7 E9 F1 F3 F5 F7 F9 G1 G3 G5 G7 G9 H1 H3 H5 H7 H9 J1 J3 J5 J7 J9 K1 K3 K5 K7 K9
D2 D4 D6 D8 D0 E2 E4 E6 E8 E0 F2 F4 F6 F8 F0 G2 G4 G6 G8 G0 H2 H4 H6 H8 H0 J2 J4 J6 J8 J0 K2 K4 K6 K8 K0
SEEBURG
WESTERN SONGS
HIT TUNES

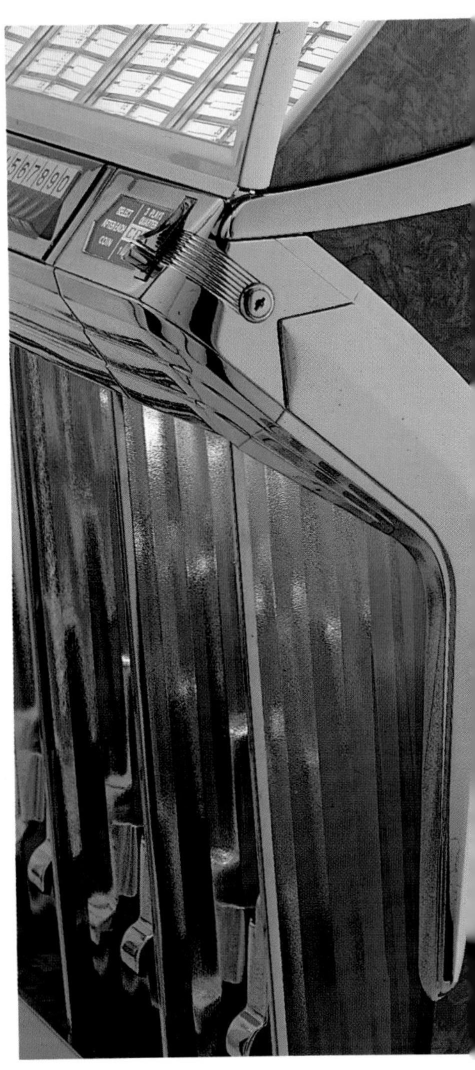

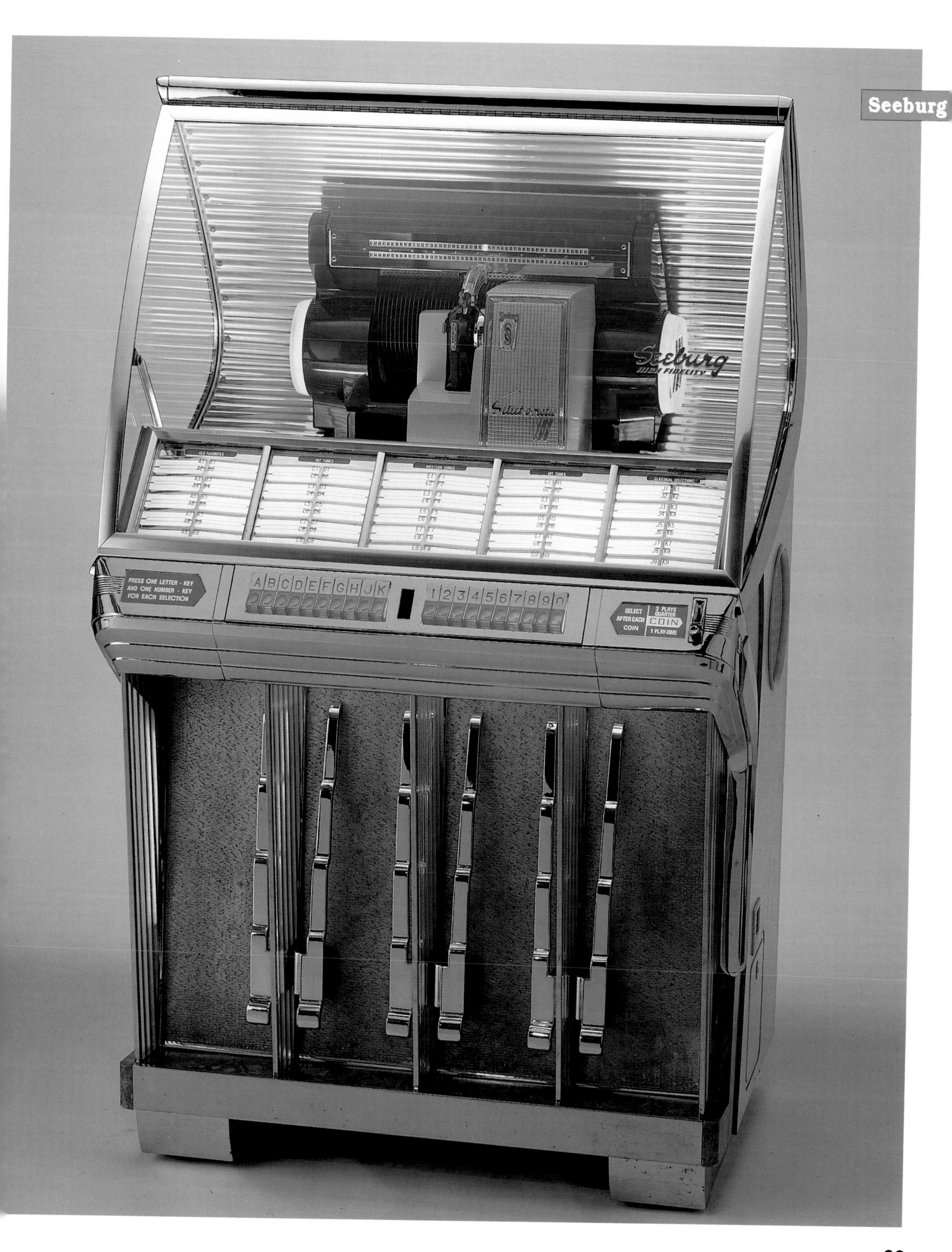
Seeburg
PRESS ONE LETTER - KEY AND ONE NUMBER - KEY FOR EACH SELECTION
A B C D E F G H J K
1 2 3 4 5 6 7 8 9 0
SELECT AFTER EACH COIN
3 PLAYS QUARTER
COIN
1 PLAY-DIME

Seeburg G

Seeburg G
Built in 1953
Made by: J. P. Seeburg Corporation,
Chicago, Illinois
Model EHF 100 G 151
Serial number 1975
100 selections

The chrome age begins at Seeburg. Two massive, plain metal columns hold the blue speaker grille between them. The light blue playing mechanism is flanked by mirrors.
The arched glass panel stretches, apparently endlessly, from the rear wall of the case to the keyboard. Side windows allow a look into the cabinet. The gray-blue artificial wood structure gives the box an elegant finish. The "45" emblems make it clear that the single record has established itself.

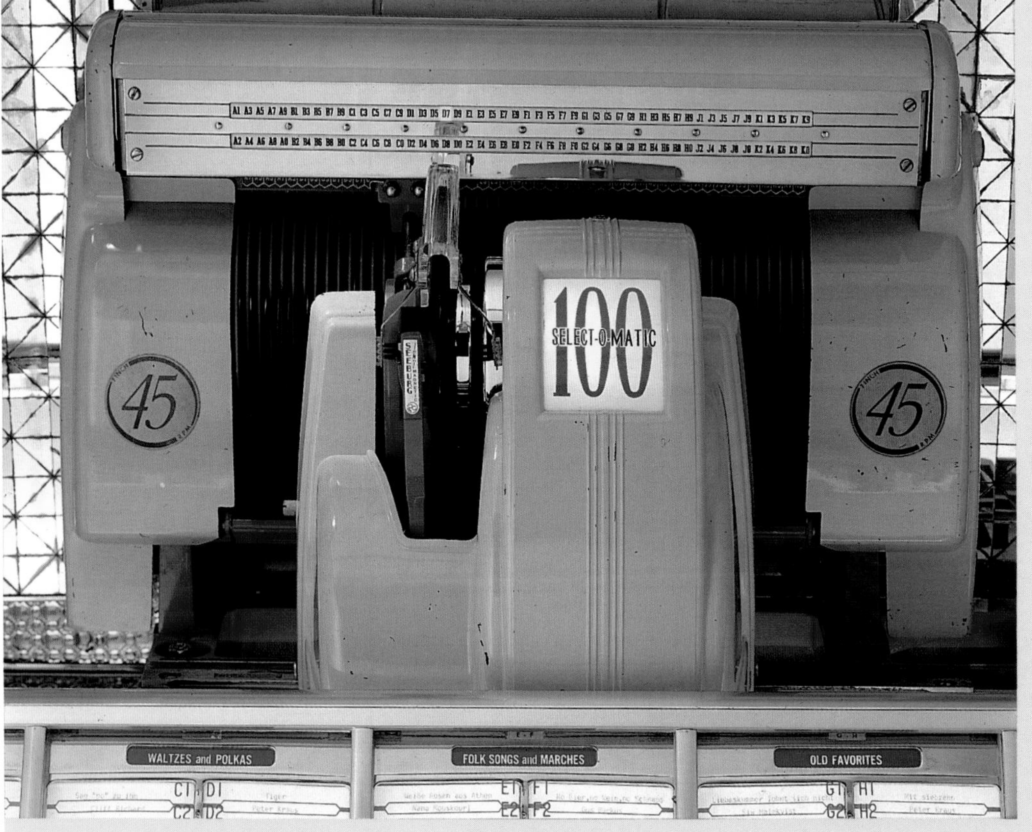

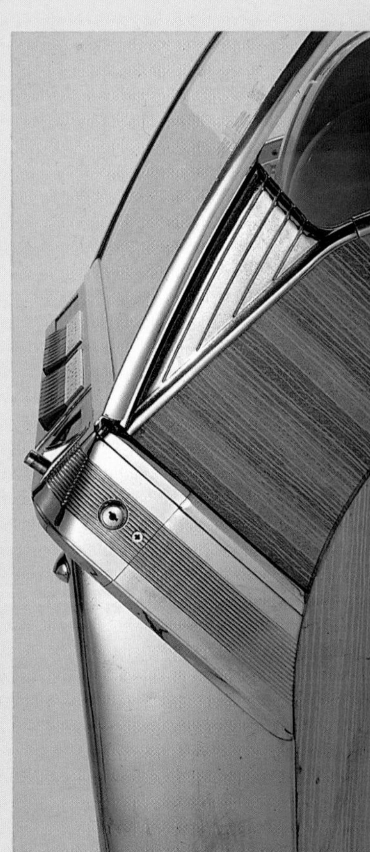

100
SELECT-O-MATIC
45
45
De Luxe
ABCDEFGHJK
1234567890
6 SPIELE - 1 MARK
MÜNZE EINWERFEN
DANN WAHL TREFFEN

Rock-Ola Princess

Rock-Ola Princess
Built in 1962
Made by: Rock-Ola Manufacturing Corporation, Chicago
Model Rock-Ola 1493 Princess
Serial number 235 958
100 selections

The final model in Rock-Ola's noteworthy design line of the fifties.
This extremely small and light floor model, developed from the two-year-older Rock-Ola 1484 wall box, affords a view of the mechanism only through a narrow, very arched glass panel.
The record magazine has been moved from the back to the left side. This layout was retained in similar form into modern times.
The design looks extremely robust and foolproof, and could usually be repaired even by laymen. The parts are nicely painted with metallic violet paint.
This model is often found in Germany, as it is reasonably priced and does not require much room.

100

SCAN
OFF
OPER.

6973
MAIN STEREO OR MACHINE
SPEAKERS POWER
25W.
12AU7
ECC82
12AX7
ECC83
6CY7
REMOTE VOL LINES
MECH.
3 AMP
240 V.
RECEIVER

AMI K

AMI K
Built in 1960
Made by: AMI Incorporated, Grand Rapids, Michigan
Model AMI JHK 200
Serial number 55 000 1
Semi-automatic selection-wheel design

In this semi-automatic version, selection is done by means of a large selection wheel with which a letter-and-number combination is set. After that, by pushing the button just to the right, the machine is started. Now the vertically mounted record magazine turns, and an iron arm puts the record on the visible turntable.
The fully automatic models were operated by big gold-colored letter and number keys.
Variations with 100 or 120 selections were also made.
The striped pattern on the sides of the box is noteworthy. All in all, this box, with its panoramic glass still in fashion at the time, makes a sturdier impression than its predecessors, the H, I and J models.

AMi
STEREO RECORDS 15¢
FAVORITES
CLASSICS
ALL TIME
YOUR PICKS
RHYTHM & BLUES
HIT TUNES
EXTENDED PLAY
POPULAR
VARIETIES
STEREO RECORDS
NUMMER EINSTELLEN 1
2 KNOPF DRÜCKEN
200 Stereophonic Selections
THE WORLD'S FINEST MUSIC

MODEL
SERIAL NO.
AMi
VOLTS 115 AC CYCLES 60 WATTS 310
WHEN WALL BOXES ARE USED ADD
100 WATTS PER TRANSFORMER
AUTOMATIC MUSIC, INC.
GRAND RAPIDS, MICHIGAN
MADE IN U.S.A.
EQUIPPED FOR OPERATION ON
220 VOLTS 50 CYCLES
AMi
GRAND RAPIDS, MICHIGAN
U.S.A.
WARNING

Wurlitzer 2410

Wurlitzer 2410
Built in 1960
Made by: The Rudolph Wurlitzer Company, North Tonawanda N.Y.
Model Wurlitzer 2410 S
Serial number 478 105
100 selections

The best-looking of the three 2400 series variations, which were available with 100, 104 and 200 selections. Aside from this technical variation, the three types differ in the sizes of their title holders. The extent of this necessarily determines how much of the mechanism and rear wall are visible.
The tone arm is styled in the shape of a stylized tail. It stretches out in front of a blue sky. The golden sun with its pattern of rays makes the beautiful day perfect. The Wurlitzer name glows on an illuminated red background. An effective combination of three colors. The big chromed speaker grille is richly illuminated.

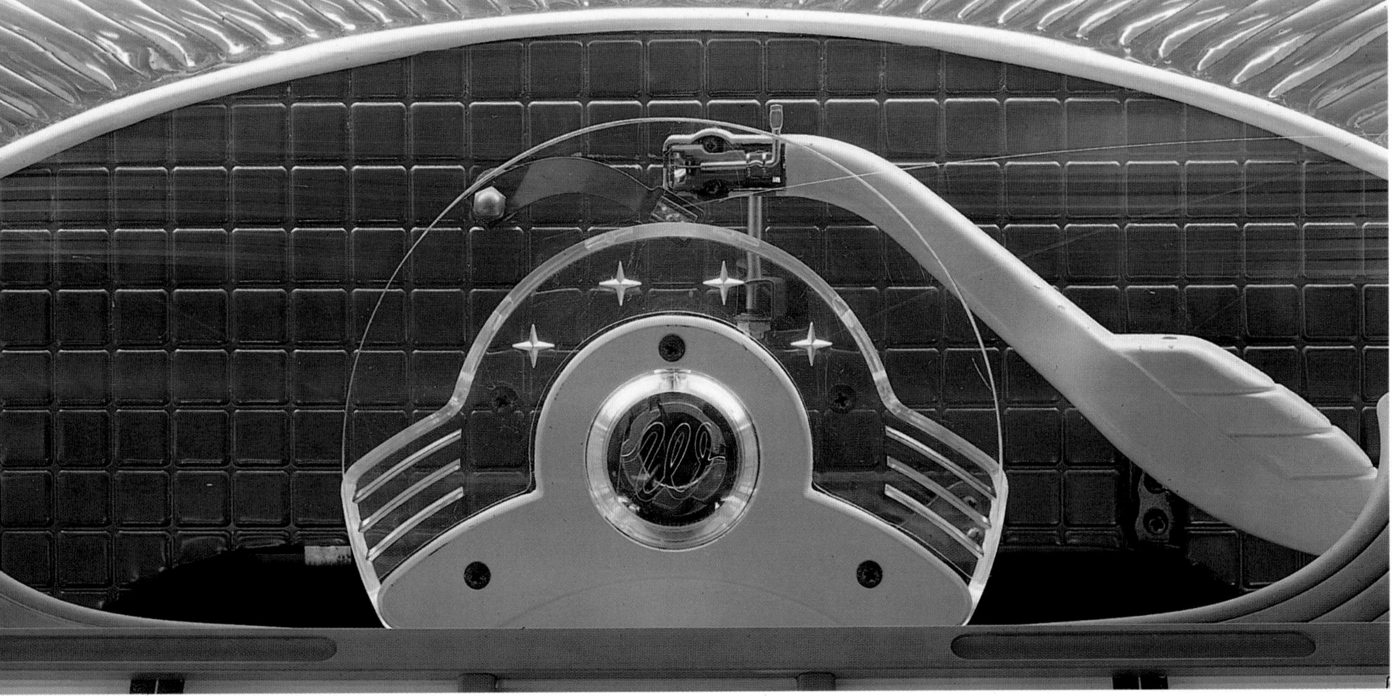

RECORD
NOW PLAYING
WURLITZER
MUSIC
PRESS A LETTER
AND NUMBER BUTTON

WURLITZER
RECORD
NOW PLAYING
WURLITZER
MUSIC
PRESS A LETTER
AND NUMBER BUTTON

Wurlitzer 241

Rock-Ola Regis

Rock-Ola Regis
Built in 1961
Made by: Rock-Ola Manufacturing Corporation, Chicago
Model Rock-Ola 1488 Regis
Serial number 222 452
120 selections

This machine actually seems older than its immediate predecessors "Tempo I" and "Tempo II." This is due in part to the somewhat old-fashioned Rock-Ola name plate on the top.
From this year on, this manufacturer also used a combination of letters and numbers for making selections.

Metallic light blue and pink tones have replaced the more vivid colors of the Tempo models. The plastic "stereo" nameplate dominates the grille.
The original advertising of 1961 showed the box beside a queen—only by chance?
Also available with 200 selections.

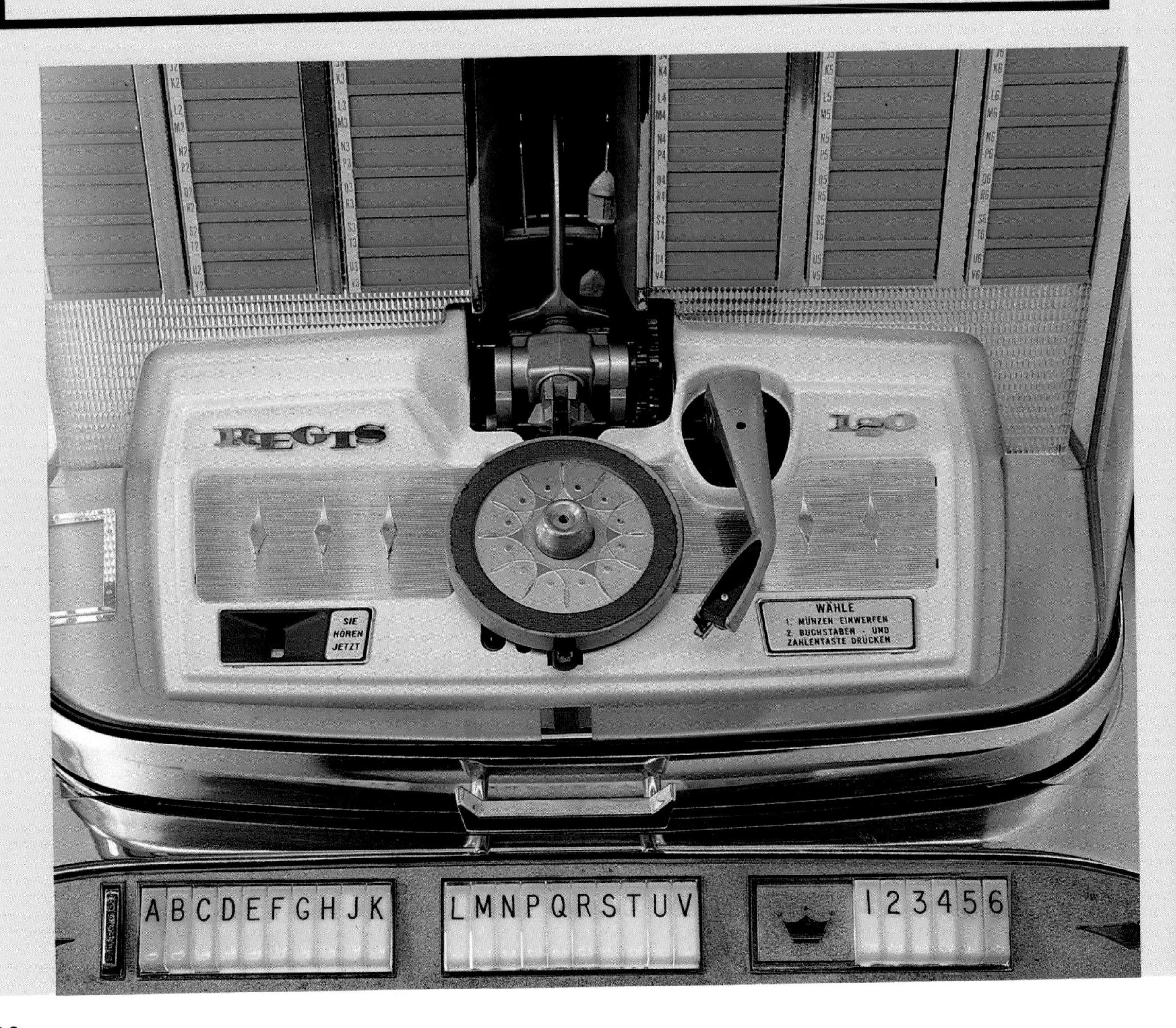

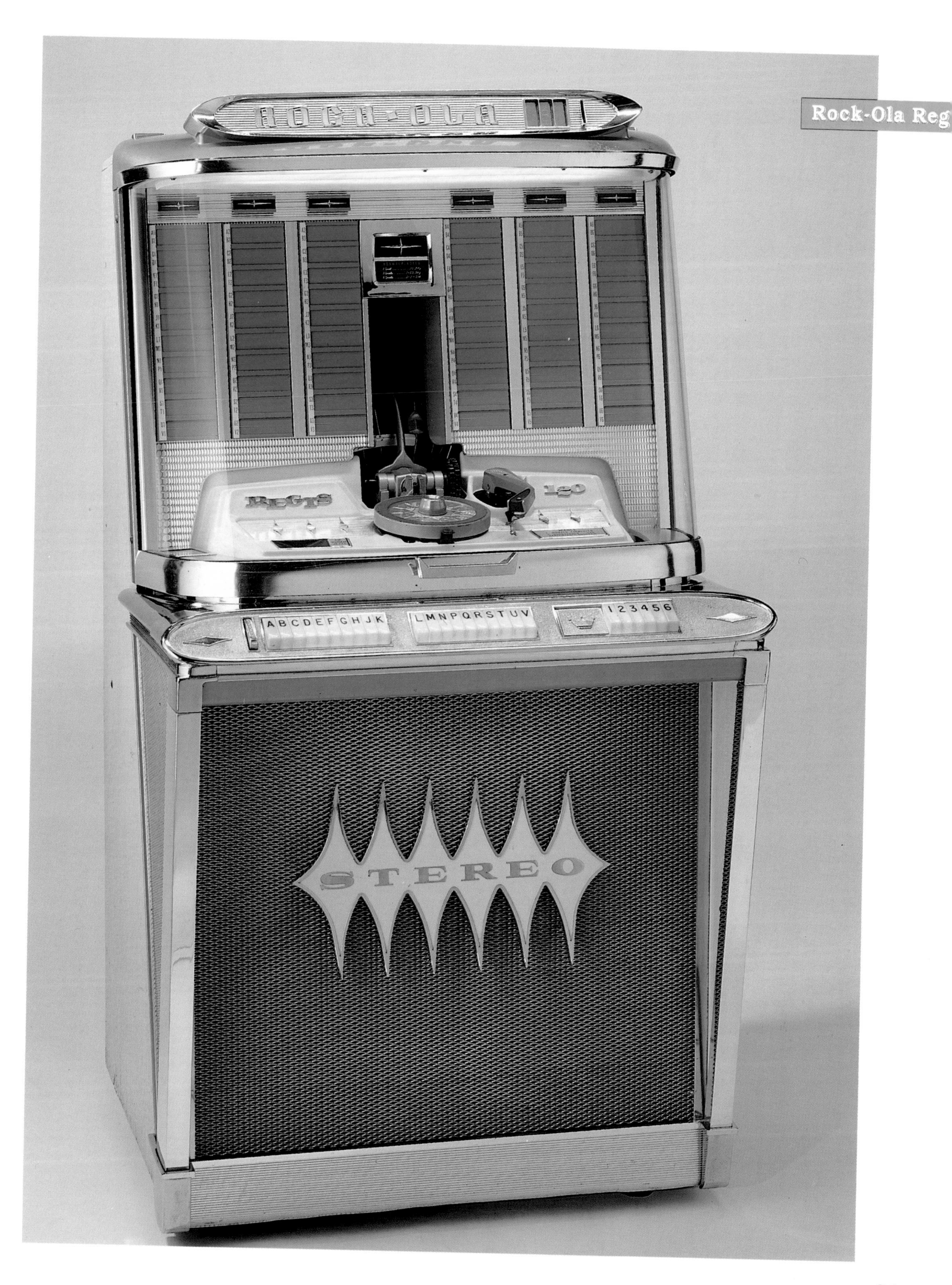
ROCK-OLA
REGIS
120
ABCDEFGHJK
LMNPQRSTUV
123456
STEREO

Rock-Ola 1454

Rock-Ola 1454
Built in 1956
Made by: Rock-Ola Manufacturing Corporation, Chicago
Model Rock-Ola 1454
Serial number 164 172
120 selections

The cabinet of the Rock-Ola 1454 of 1956 looks as if it were wrapped in cellophane with a star design. Scarcely changed technically from its 1448 predecessor, the box has changed its colors like a chameleon. Cold turns warm as silver turns to gold. The solid-color sides of the case have been divided into two differently colored areas by a diagonal dividing line, with a marbleized field of brown-beige or green-beige above and a simple imitation wood below.
Until 1958 there were still several variations of this series, which differed in design.
After that, the Tempo series provided a completely new appearance.

ROCK·OLA
PROGRAMM-
WECHSEL
MÜNZ-
EINWURF
120 Hi-Fi Selections
ROCK·OLA

120 Hi-Fi Selections

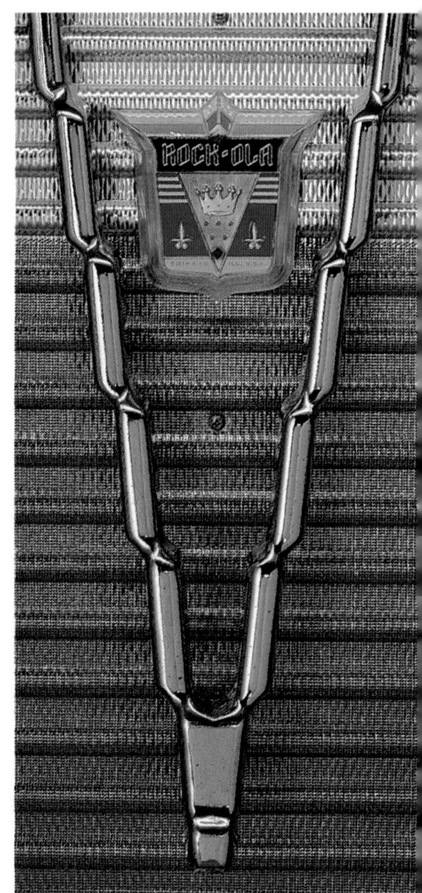
ROCK-OLA

Wurlitzer 1800

Wurlitzer 1800
Built in 1955
Made by: The Rudolph Wurlitzer Company, North Tonawanda, N.Y.
Model 1800
Serial number 000 165 1
104 selections
Made in Germany by Automatenbau Gustav Husemann

The circular horizontal record magazine holds 52 records (104 selections). A lifting mechanism takes the selected record and brings it to its playing position. The vertical playing position is typical of Wurlitzer machines. Selection is made by a letter-and-number combination, with letters A, B, C and D and numbers 1 to 26.
The wavy bright-red background can also be found in the interior decor of contemporary bars and cinemas. The arched side panels are unique.

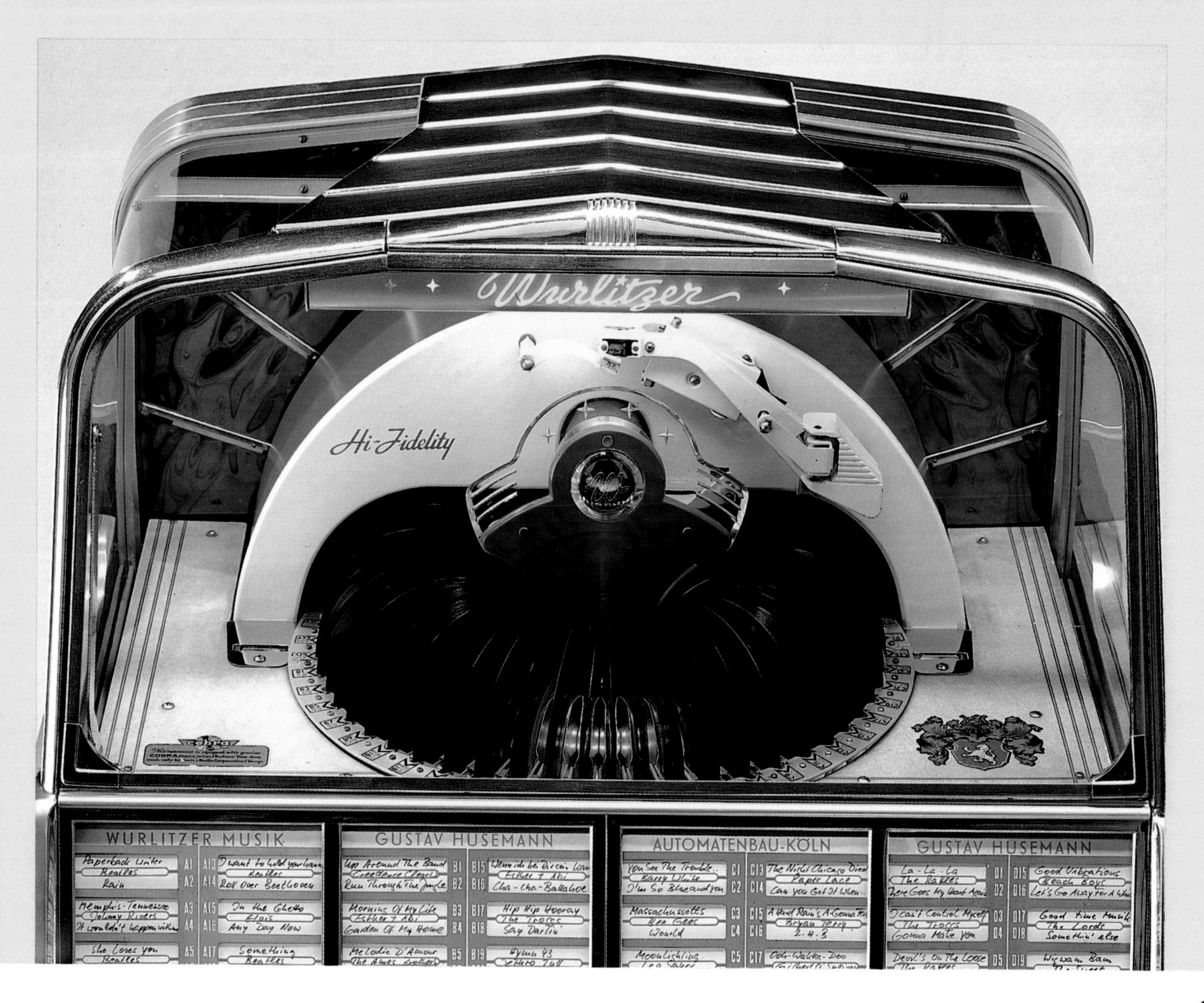

Wurlitzer 1800

AMI Continental

AMI Continental
Built in 1961
Made by: AMI Incorporated, Grand Rapids, Michigan
Model XJCA-200
Serial number 596 291
200 selections
Fully automatic keyboard

The spacecraft of the time inspired this juke-box design.
The record plays under a glass bubble, the title labels are found on a stylized parabolic mirror.
Advertising promised the buyer "a design for tomorrow for a good return today."
The AMI Continental is one of the very few jukeboxes that was also styled on the back and thus could stand free in a room.

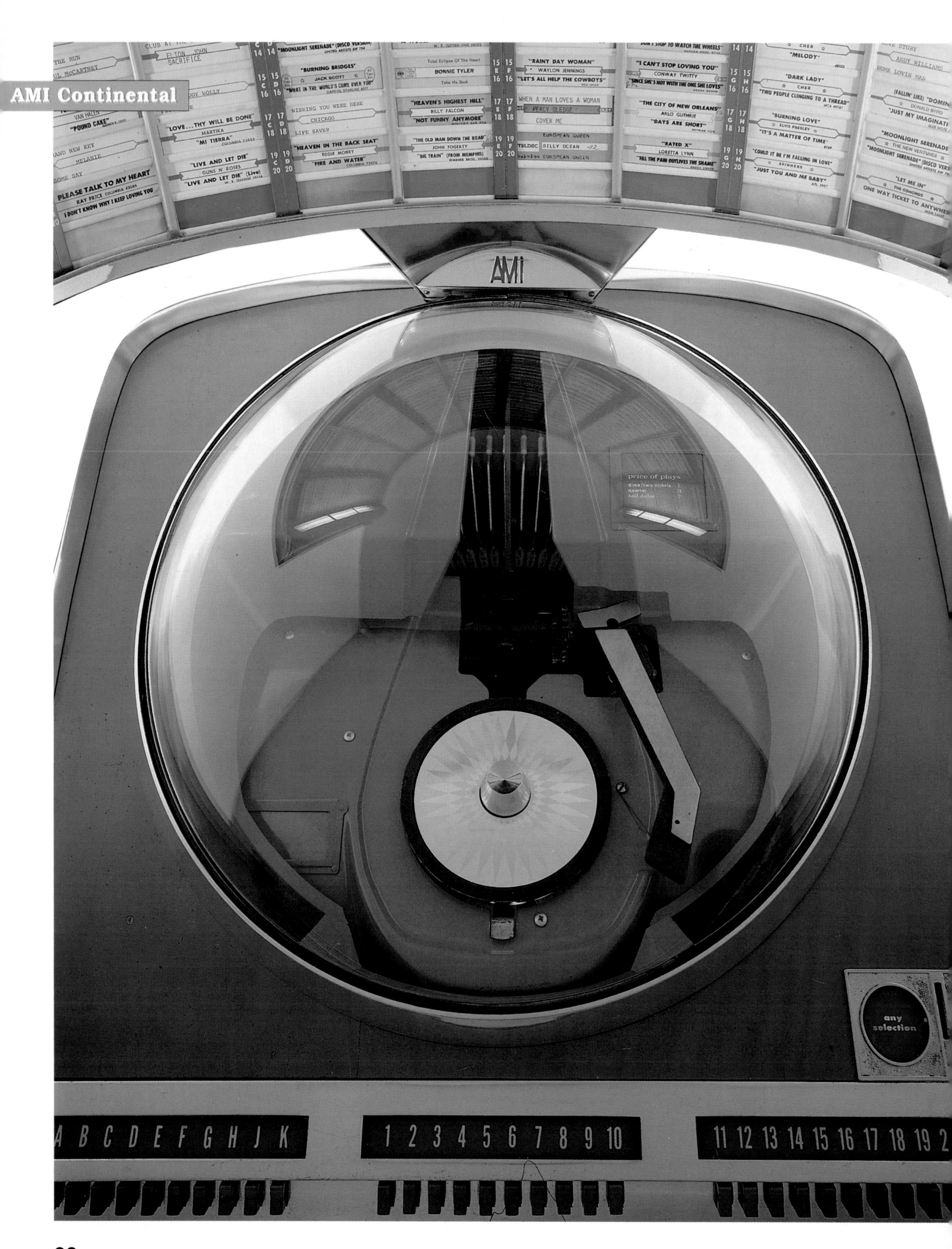
"MOONLIGHT SERENADE" (DISCO VERSION)
"BURNING BRIDGES"
JACK SCOTT
"WHAT IN THE WORLD'S COME OVER YOU"
Total Eclipse Of The Heart
BONNIE TYLER
Take Me Back
"RAINY DAY WOMAN"
WAYLON JENNINGS
"LET'S ALL HELP THE COWBOYS"
"I CAN'T STOP LOVING YOU"
CONWAY TWITTY
"MELODY"
"DARK LADY"
CHER
"TWO PEOPLE CLINGING TO A THREAD"
ELTON JOHN
SACRIFICE
"POUND CAKE"
VAN HALEN
"LOVE... THY WILL BE DONE"
MARTIKA
"MI TIERRA"
WISHING YOU WERE HERE
CHICAGO
LIFE SAVER
"HEAVEN'S HIGHEST HILL"
BILLY FALCON
"NOT FUNNY ANYMORE"
WHEN A MAN LOVES A WOMAN
PERCY SLEDGE
COVER ME
"THE CITY OF NEW ORLEANS"
ARLO GUTHRIE
"DAYS ARE SHORT"
"BURNING LOVE"
ELVIS PRESLEY
"IT'S A MATTER OF TIME"
ANDY WILLIAMS
HOME LOVIN MAN
"JUST MY IMAGINATI
"LIVE AND LET DIE"
GUNS N' ROSES
"LIVE AND LET DIE" (Live)
"HEAVEN IN THE BACK SEAT"
EDDIE MONEY
"FIRE AND WATER"
"THE OLD MAN DOWN THE ROAD"
JOHN FOGERTY
"BIG TRAIN" (FROM MEMPHIS)
EUROPEAN QUEEN
BILLY OCEAN
"RATED X"
LORETTA LYNN
"TILL THE PAIN OUTLIVES THE SHAME"
"COULD IT BE I'M FALLING IN LOVE"
SPINNERS
"JUST YOU AND ME BABY"
"MOONLIGHT SERENADE"
THE NEW VENTURES
"LET ME IN"
THE OSMONDS
ONE WAY TICKET TO ANYWHER
PLEASE TALK TO MY HEART
RAY PRICE
I DON'T KNOW WHY I KEEP LOVING YOU
MELANIE
AMI
price of plays
dime/two nickels
quarter
half-dollar
any selection
A B C D E F G H J K
1 2 3 4 5 6 7 8 9 10
11 12 13 14 15 16 17 18 19

AMi
Continental

Panoramic

Panoramic
Built in 1959
Made by: Tonomat-Automaten, Neu-Isenburg, Germany
Model Panoramic 200
Serial number 22 649
200 selections

The big panoramic pane of plexiglas gives a good view of the playing mechanism. Advertising promised "astoundingly simple and pleasant pre-play selection with a new type of program carousel. Read the program, make a direct choice of the desired title in the window, and then push the button—that is all."

TONOMAT

Schlager des Monats
Tanzmusik
Bedienungs-Anleitung

Seeburg 201

Seeburg 201
Built in 1958
Made by: J. P. Seeburg Corporation,
Chicago, Illinois
Model 201
Serial number 1679
200 selections
Made under license in Germany by
Löwen-Automaten

One of the three models with a Cadillac-style design, with red lights on the speaker grille. Strictly speaking, they do not come from the Cadillac, but from rockets and jet planes, which influenced both automobile and juke-box design.
From 1957 on, Seeburg did away with arched glass panels, beginning this series with the KD 200.
The moving record player takes its position in front of the chosen record in the straight-line record magazine. A gripper moves the record into playing position.

MUSIK FÜR ALLE
SELECT-O-MATIC
Two Hundred
Seeburg
A B C D E F G H J K
1 2 3 4 5 6 7 8 9 0
L M N P Q R S T U V

Stufenschalter-Kontakte
Kontakt-spiel in mm
Kontakt-druck in g
Normal-Stellung
Zeit-Relais 1
Zeit-relais 2
Transport-Relais
Nummern-Stufen-Magnet
Buchstaben-Stufen-Magnet
abschrauben
Ankerluft:
Zeitrelais 1: 2,4 mm
Transportrelais: 1,2 mm
Zeitrelais 2: 2,4 mm
WARNING
TURN OFF MAIN POWER
SWITCH BEFORE REMOVING
J510
J505
GND
70 VOLT
C.V.
SPEAKER
J103
PHONO
SPEAKERS
J102
REMOTE VOLUME
CONTROL
12AX7
XV104
PHASE SPLIT
AND DRIVER
SEEBURG
HIGH FIDELITY
MASTER AMPLIFIER
TYPE HFMA2
CODE B
117 VOLTS
115 WATTS
THE SEEBURG
CORPORATION
POWER OUTPUT
F101
2 AMP
SLO-BLO
FUSE
ROOM CONDITIONS
BASS SETTING
ACOUSTICALLY LIVE 3 or 4 3 or 4
AVERAGE 2 or 3 4
ACOUSTICALLY DEAD 1 or 2 5 or 6
NOTE:
GOOD QUALITY REPRODUCTION CANNOT BE OBTAINED WITH
POOR QUALITY RECORDS OR WITH WORN STYLUSES. PLEASE
REMEMBER — WORN STYLUSES WILL DAMAGE YOUR RECORDS.
20W
10W
4W
1W
SPEAKER TEST
TIEFTÖNE
HOCHTÖNE
6BJ6
XV102
A.V.C. CONTROL
12AX7
XV103
TONE CONTROL
AMPLIFIER
CATHODE FOLLOWER
12AX7
XV101
A.V.C. CONTROL
AMPLIFIER
J104
MUTE
SQUELCH
J101
AUDIO
INPUT
Achtung!

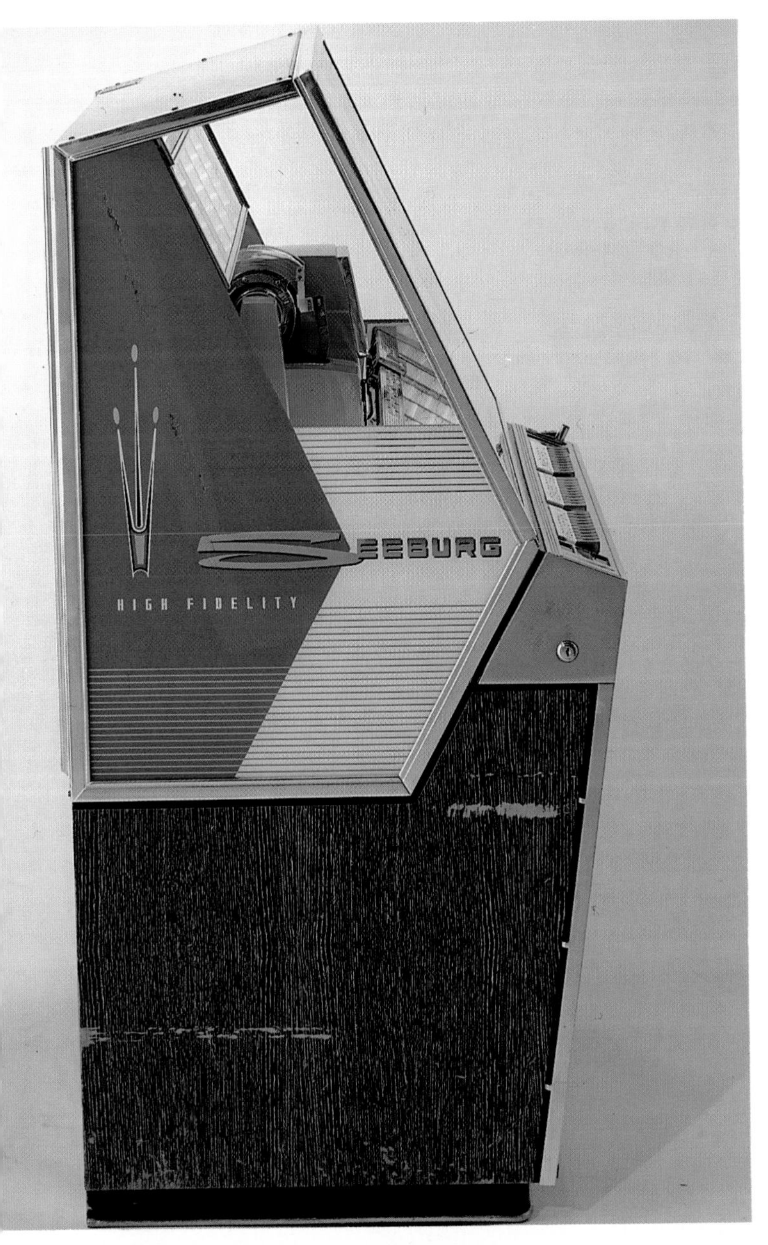
SEEBURG
HIGH FIDELITY

Seeburg Höchste Tontreue
Select-o-matic 200
Modell 201
Geräte-Nr. 1579
Volt 220
Per. 50
Stromverbrauch im Ruhezustand: während des Betriebes:
125 Watt 0.7 Amp. 260 Watt 1.5 Amp.
Generalimporteur für Deutschland: LÖWEN-AUTOMATEN
BINGEN (RHEIN) Tel. 5144 FS 042-889
EIGENE NIEDERLASSUNGEN:
DÜSSELDORF Tel. 49 14 11
HAMBURG Tel. 43 71 47
BRAUNSCHWEIG Tel. 309 21

SEEBURG
SELECT-O-MATIC

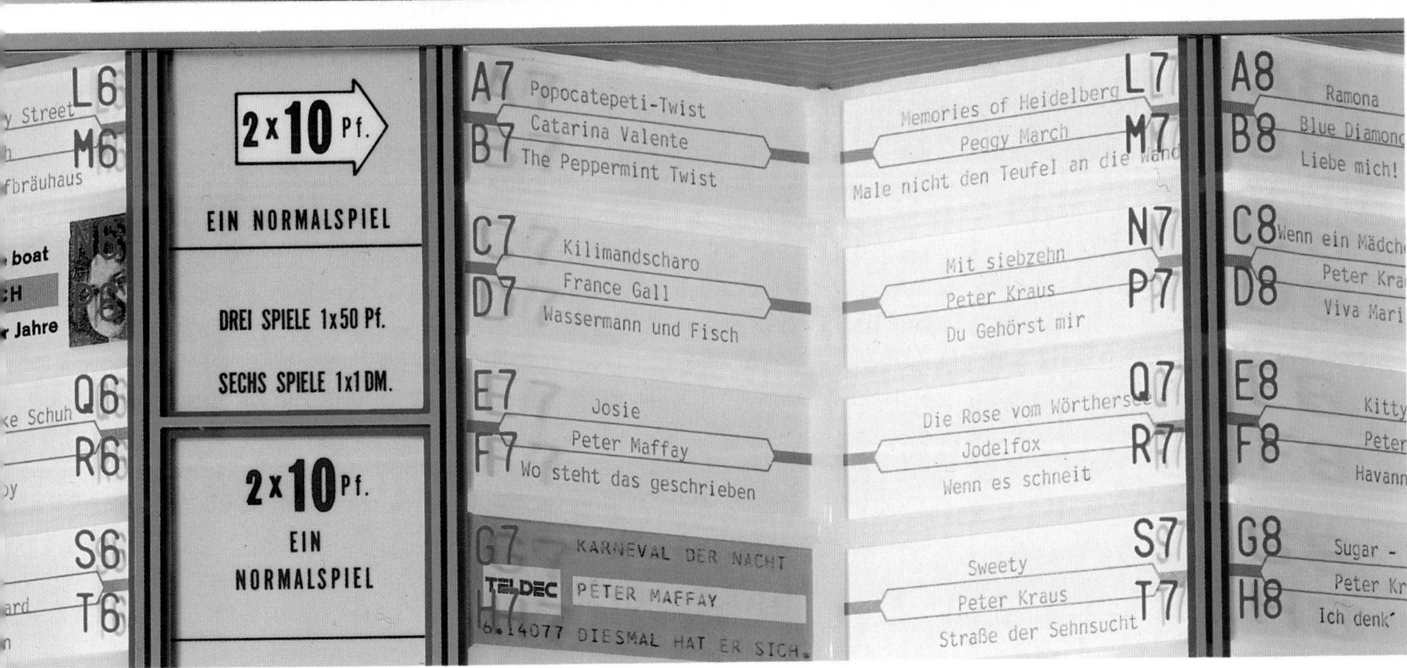
L6
M6
Q6
R6
S6
T6
2 x 10 Pf.
EIN NORMALSPIEL
DREI SPIELE 1x50 Pf.
SECHS SPIELE 1x1 DM.
2 x 10 Pf.
EIN NORMALSPIEL
A7 Popocatepeti-Twist
Catarina Valente
B7 The Peppermint Twist
C7 Kilimandscharo
France Gall
D7 Wassermann und Fisch
E7 Josie
Peter Maffay
F7 Wo steht das geschrieben
G7 KARNEVAL DER NACHT
TELDEC
H7 PETER MAFFAY
6.14077 DIESMAL HAT ER SICH
L7 Memories of Heidelberg
Peggy March
M7 Male nicht den Teufel an die Wand
N7 Mit siebzehn
Peter Kraus
P7 Du Gehörst mir
Q7 Die Rose vom Wörthersee
Jodelfox
R7 Wenn es schneit
S7 Sweety
Peter Kraus
T7 Straße der Sehnsucht
A8 Ramona
B8 Blue Diamond
Liebe mich!
C8
D8
E8
F8
G8
H8

Rock-Ola Tempo II

Rock-Ola Tempo II
Built in 1960
Made by: Rock-Ola Manufacturing Corporation, Chicago
Model Rock-Ola 1485 Tempo II
Serial number 212 597
200 selections

The second number in the four-part series from Tempo to Empress is especially striking on account of its sporting elegance.
The blue-and-cream sail on the black-and-silver speaker grille, almost like a surfing logo, makes it unmistakable in Rock-Ola's long history of jukeboxes. Under the big panoramic glass panel, the title labels are set in a folded arrangement. The five selection buttons each represent forty selections. They are reminiscent of a piano.—The Cadillac-style fins, already familiar from the "Tempo" forerunner, are relics of a crazy era. Also available with 120 selections.

WÄHLE
Tempo II
200
7 8 9 10 11 12 13 14 15 16 17 18 19 20 21 22 23 24 25 26 27 28 29 30 31 32 33 34 35 36 37 38 39

200

Seeburg VL

Seeburg VL
Built in 1956
Made by: J. P. Seeburg Corporation,
Chicago, Illinois
Model: label missing
200 selections

The battleship among the jukeboxes in the style of the fifties: much bigger, much heavier than its forerunners.
For the first time in the world, Seeburg offered, in the Model V of 1955, a jukebox with 100 records (200 selections). The successor "VL" model of 1956—differing only by its colors—was just as successful on the market.
The gigantic numbers of title labels were hidden on a program roller. It was set in motion by pre-choice keys and thus afforded a look at individual areas such as "Rhythm and Blues" and "Rock 'n Roll."
Seeburg's engineers were particularly proud of the newly developed, service-free "Select-O-Matic" magnetic selection mechanism, as well as the "Dual-Price-System."

200
SELECT-O-MATIC
Seeburg
Two Hundred Selections
High Fidelity
Hit Tunes
Seeburg

200
SELECT-O-MATIC
Seeburg

Seeburg

Seeburg J

Seeburg J
Built in 1956
First built in America in 1955
Made by: J. P. Seeburg Corporation,
Chicago, Illinois
Model HF 100 J
Serial number 2466
100 selections

Developed directly from the predecessor Model R, the J continued the high standard of Seeburg machines in that decade. The German model, built under license by Löwen-Automaten, did not differ at all—except for the nameplate—from the American version. Even the screws are the same type.
In front of the white record magazine there moves the typical Seeburg playing mechanism—hidden under a yellow and red plastic cover. The speaker area is also a new design. A covering nicknamed "bathmat" today protects the speaker grid from being kicked too hard. At the same time as the Seeburg J, the first 200-selection box, the Seeburg V 200, appeared in America.

100
Seeburg

Remote Controls

The lovingly styled remote-control selection boxes of the fifties and early sixties appear to the beholder like miniature jukeboxes. They are known as wall boxes. Cafe and restaurant owners liked to use these important auxiliaries in order to spare their visitors the "laborious" path to the jukebox.
Mounted on tables and bars, they competed for the guest's favor. Connected to the box by cable, they had their own keyboard with which any record could be chosen.
The title labels were mounted on turning pages, in book form, to save space. Coin slots and sorters were also housed in these devices, but not the controls such as the speaker volume regulator or shutoff knob. Only the proprietor could control them.
For particularly small places, all the manufacturers also offered so-called "hideaway" jukeboxes at that time. As the name indicates, these boxes had to disappear from the field of vision: simple wooden cases with record magazine and complete playing technology, but without keyboard, title holder or speaker.
Activated by remote control, they played the music into the place from their hiding place. No guest ever got a look at them. The term "wallbox," translated into the German "Wandbox," often led to misunderstanding.
At this time, various firms manufactured jukeboxes that were mounted on the wall to save space, and so did not stand on the floor. Necessarily, these devices are clearly larger than the remote controls, for they also had to house the record magazine and playing mechanism.

Seeburg Wall-O-Matic 3W1 (since 1948)

Seeburg Wall-O-Matic 160, S3W (since 1958)

Hecker revision of Seeburg 3W1 (sixties)

Seeburg Wall-O-Matic 200, V-3WA (1956)

AMI W 120 (1953)

Rock-Ola 1555 120/200 (1959)

AMI 2000, WQ-200 (1956)

Wurlitzer 5250 (1961)

The Stars of the Fifties—Records and Successes

The world was torn apart in the fifties in two senses: divided into east and west, split into young and old.

For one thing, there was the tense political situation: two superpowers, allies a short time before in a common war against Germany, were now implacably opposed. The world was divided into capitalism and socialism, the Soviet Union and the USA became bitter enemies in the Cold War. This "Cold War" became a concept that was to define the political situation in the next decades.

Nobody could afford to ignore it, and thus a genuine race began for the bomb, after America had become the first country to develop and use the atomic bomb—on Hiroshima and Nagasaki in 1945. In 1949 the Soviets were also capable of building an atomic bomb—and both superpowers immediately concentrated on the escalation of everything—the hydrogen bomb, the ultimate superbomb, was developed and tested by the USA in 1952 and in the very next year, 1953, by the USSR. For the first time in human history there was a permanent danger that not merely one country but the entire globe would be annihilated in a war. The concept of "overkill" describes the weapon's potential: The world could be destroyed several times over.

In the periodical "Magnum" there appears an article in Issue 24 of 1959 called "Life After 45" by editor Karl Pawek, in which he describes the situation at that time: "We build secure palaces, as if they were intended to last a thousand years—but the tactical bomber squadrons are already in the air. Their deployment is not a question of years or months, but of minutes."

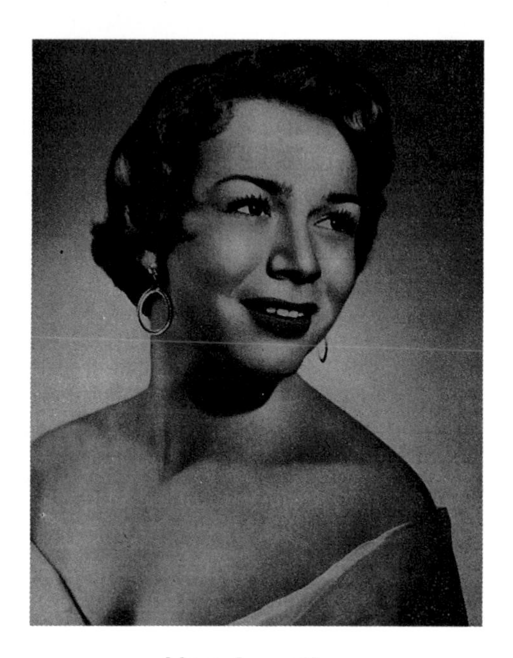

Liane Augustin

In Search of a Healthy World

As ominous as the political situation was—the unthinkable, the annihilation of all life, could take place at any time—it was a good time for people in a material sense. In 1950 most food rationing was ended.

The constant fear of the bomb and overkill on the one hand and the good life on the other, moved some people, especially young people, to take on a critical attitude to the world and the older generation. At first they rebelled in vague, untargeted ways; later they became politically aware and had definite goals.

The star singers were no longer the same, and they sang songs not so much about deep feelings as about shallow pleasures. The pop song, the gently lulling song without deeper meaning, relieved the people of the necessity of thinking seriously, giving them a short escape.

This becomes apparent in the very titles. Bill Ramsey sang significantly of a chocolate ice-cream vendor:

Bibi Johns and Peter Alexander making a movie featuring one of their jukebox successes. (1957)

Yes, the man came to me out of thin air,
And he had only one eye and one horn,
He pushed a cart, and he looked as if
He were at home up on Mars.
It was the Wumba Tumba chocolate ice-cream vendor
From another star.

Another big name from this era was Peter Alexander, whose titles, among many others, include "I know what you lack," "Mandolins and Moonlight," "Bambina," and "Ciao, Ciao, Bambina."

Here are words from the song "The Color of Love," written by Ralph Siegel, interpreted by Paul Kuhn:

Instead of white, wear red,
That is the color of love.
And the man already knows:
Wear blue instead of green.
That is the color of loyalty,
Then your clothes speak for you.

If you are jealous,
Wear yellow when he kisses you
Or put on lilac,
So he cannot suspect.

Instead of white, wear red,
That is the color of love,
Every man knows that!

Of course, these words sound better when sung.

Cornelia Froboess and Peter Kraus

Gerhard Wendland

Vico Torriani

Fred Bertelmann

The stars and starlets of those days can be regarded to this day as a "Who's Who" of the pop-music industry; they have survived in show business or at least enjoy a momentary comeback.

One can see what many people wanted to hear at that time, as in the Rodgers-Duo's title, "Be Satisfied With Today."

Caterina Valente and Bill Haley. Bill Haley made two movies, released in 1956: "Rock Around the Clock" and "Don't Knock the Rock."

Top Hits of the Month

				Last Month
1.	**Banana Boat**	Harry Belafonte	RCA-Victor 47-6711	1
2.	**Homeless**	Freddy	Polydor 23,381	2
3.	**Charcoal Lisa**	Heimatsänger	Decca 18,539	-
4.	**One More Ice Cream**	Peter Alexander	Polydor 23,400	-
5.	**Little Bimmelbahn**	Harz Yodelers	Metronome DM45-24	-
6.	**Why do the Stars Shine Tonight**	Wolfgang Sauer	Electrola 17-8667	4
7.	**Tipitipitipso**	Caterina Valente	Polydor 23-403	5
8.	**Cindy, oh Cindy**	Margot Eskins	Polydor 23,363	3
9.	**Mi Casa, su Casa**	Perry Como	RCA-Victor 47-6815	-
10.	**I did it all for Love**	Peter Alexander	Polydor 23,432	-

Harry Belafonte

Only with the appearance of Caterina Valente and Bill Haley together in the film "Hier bin ich" (Here I Am) did Rock 'n Roll become accepted by the establishment in Germany. But then came new musical directions that were significant for young people. The Beatles and the Rolling Stones were there to ring in the sixties.

A Time of Rebellion

In the fifties, though, the young wanted something different, including hearing different music. What was being played in the jukeboxes was, as a rule, very different from what was being sold in record shops.

This duality of society was reflected in the musical tastes of the time. On the one hand, there was the bright world of Peter Alexander in the jukeboxes; on the other were Elvis Presley, Bill Haley and all the others who had something to express in their songs—even if it was just the general attitude of rejecting the old ways. The young people felt alone, completely dissatisfied, disillusioned and without models. They found themselves in a very different kind of music—Rock 'n Roll.

The once-ignored music of the black community from the South and the ghettos, the blues, found its way into other branches of music. Only at the end of the forties was the concept of "Rhythm & Blues" created. Then Rock 'n Roll picked up these elements. The words became more direct. Bill Haley and Elvis Presley were among the first stars of Rock 'n Roll, though followed immediately by black artists such as Chuck Berry, Fats Domino and Little Richard. In the film "Seed of Power," and "Rock Around the Clock" in 1956, a whole generation found itself described and was thrilled. Tumultuous scenes took place, whole cinemas were smashed.

Youth sought—and found—its idols outside the adult world. One such idol was James Dean. The young James Dean (1931-1955) became world-famous in eighteen months—with only three films from Hollywood's dream factory: "East of Eden," "Rebel Without A Cause" and "Giant." He died on September 30, 1955, in an auto accident in his silver-gray Porsche Spyder.

At that time, the film played a much greater role in public awareness and had much more influence than is the case today. In the fifties there was a whole row of outstanding films. Here are only a few examples:

1951
A Streetcar Named Desire. Director Elia Kazan, stars Vivien Leigh and Marlon Brando.

James Dean

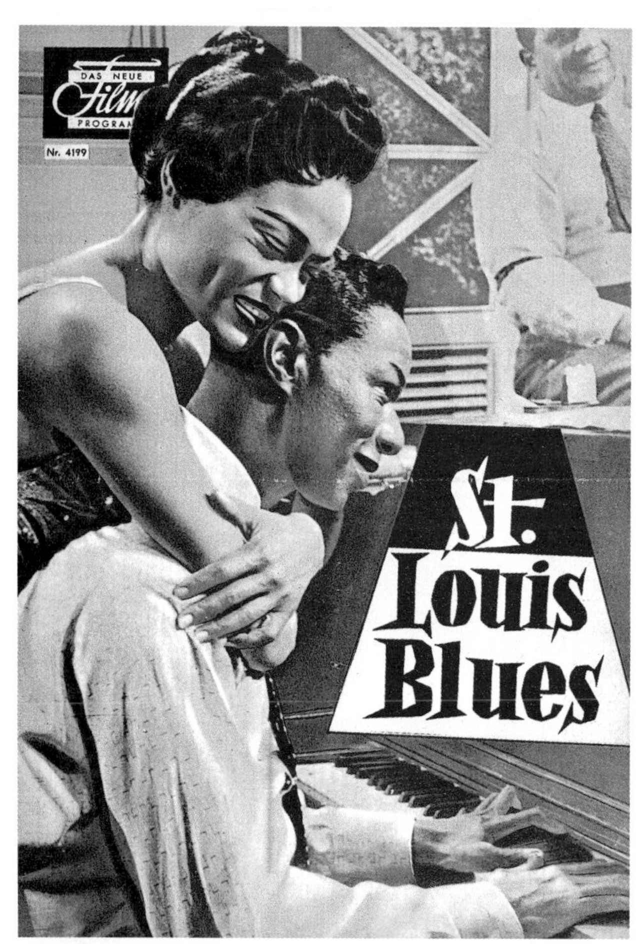
DAS NEUE
Nr. 4199
St. Louis Blues

Illustrierte
Film-Bühne
Nr. 3708
Die oberen
(HIGH SOCIETY)
ZEHNTAUSEND
W
EIN FARBFILM IN VISTAVISION

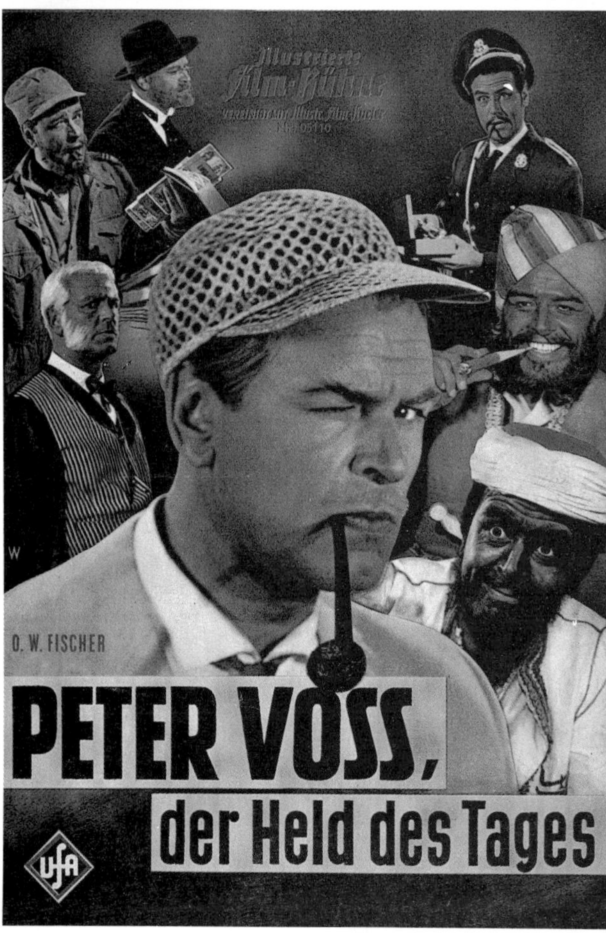
W
O. W. FISCHER
PETER VOSS,
der Held des Tages
UFA

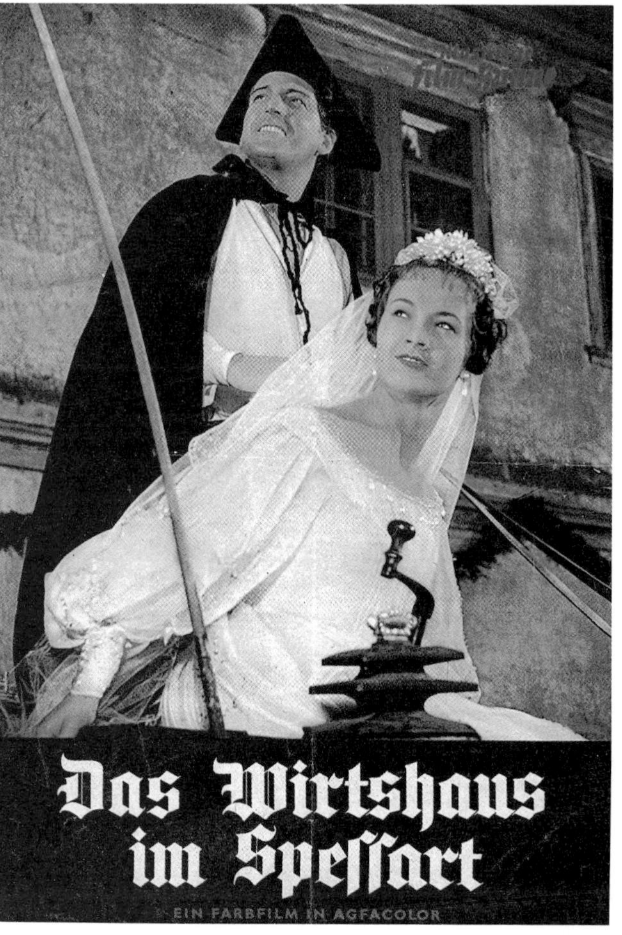
Das Wirtshaus
im Spessart
EIN FARBFILM IN AGFACOLOR

1953
From Here to Eternity. Director Fred Zinnemann, stars Burt Lancaster and Frank Sinatra.

1955
East of Eden. Director Elia Kazan, star James Dean.

1957
The Prince and the Showgirl. Director Laurence Olivier, star Marilyn Monroe.

1958
Breathless. Director Jean-Luc Godard, star Jean-Paul Belmondo.

1960
Misfits. Director John Huston, stars Marilyn Monroe, Montgomery Clift and Clark Gable.

This was the environment in which Rock 'n Roll was heard and young people were moved to hitherto unknown emotions. A quote from "The Fifties" by Paul Maenz says: "Half-grown girls wet the cinema seats, which were then destroyed by the aggression of their boyfriends...In front of their hangouts, packs of motorcycles were parked. Inside taverns, furnished with the cheap effects of rumpus-room romanticism, vibrated to the rock numbers of Bill Haley, Elvis Presley and Little Richard that thundered from the jukeboxes."

For us today it is scarcely conceivable what a revolution Elvis Presley (1935-1977) amounted to then. His swinging hips, his movements while he sang, had a truly demanding sexuality; his voice was not silky-soft, but provocative and fresh. With him and Rock 'n Roll on the scene, the grand masters of popular music, i.e. Frank Sinatra, suddenly belonged to the old folks.

Musical films, like those in the program on page 110, offered everything from blues and jazz to a touch of folk music.

Elvis symbolized animalism and moved a whole generation of young people, who idolized him like a god for it.

The eruption of a young generation that did not know where it wanted to go but knew what it did not want any more, led to unrest and a touch of violence: gangs came into being. Only later—in the sixties—did this dissatisfaction develop into politically oriented awareness. The teenagers of those days were the forerunners and precursors of a rebellious generation that soon began to make people concerned about more than just brawling: the student revolts of the late sixties cannot be explained without the rockers of the fifties.

For the first time, it was not a matter of the young people devoutly following the adults and taking over their idols. They wanted, clearly and decisively, to separate and delineate themselves apart from the bourgeoisie.

Pegged pants, V-shaped jackets, padded shoulders, shirts with stubby collars and ties as thin as strings were the preferred apparel of the young. They combed their hair with a wave in front and a ducktail in back.

The girls wore stiffly starched skirts and petticoats and scarcely knew where they could sit down. The wide sweater with an asymmetrical shawl collar and three-quarter-length slacks (wide on top, narrow at the hips with little refined slits), plus flat ballerina shoes were also very popular pieces of clothing. The young had their own music and also differed from grown-ups in terms of clothing.

Collecting Jukeboxes

"My life is rhythm," says Elvis Presley on an old movie placard in the home of Petra Reutter in the Ruhr area.

Her enthusiasm and involvement are dedicated to jukeboxes that can play his songs well.

Jürgen Lukas talks with the collector, who formerly worked as a press photographer and now is occupied mainly as a mother:

Jürgen: I have known your family now for several years, and whenever we got together, it was because of the hobby we share: the jukeboxes, Petra.. You and your husband Willi have one of the nicest and most extensive private collections in Europe. For that reason, at your house I have probably come to the right place to inflict a hundred questions on you as a reporter for our readers—since your collection is making up the pictorial section of this book. But first, tell me, in your eyes, what is a collector?

Petra: That is a person who is intensively occupied with a subject and assembles various specimens of it which then, as a whole, form an expressive unity.

In reference to jukeboxes, that means either that one collects according to eras of style, different technical aspects, or different brands. But there could be other criteria too.

Jürgen: In which you have decided on a certain period of time, that of the 45 rpm single which means, if you want, from the early fifties to the late eighties.

A lineup of jukeboxes, such as here in Wieze, Belgium, in 1990, as internationally operating dealers bring boxes from all over the world together. Their work as breadwinners is far behind them. But as collectable design objects, they do not belong—like most from their times—in the scrap heap, but in the good hands of fans.

Golden Age—Silver Age

Petra: That is absolutely right. We are only interested in the boxes that are suitable for playing the old single records. And this is actually not for technical reasons, but rather on account of their appearance. For the breakthrough in technology took place simultaneously with the breakthrough in design.
The boxes for shellac records have also interested us, of course. But they dropped out of the picture at some time.
Jürgen: But for just these devices the Americans use the label "Golden Age." The boxes of the fifties only qualify for the label "Silver Age." How do you feel about that?
Petra: On the basis of the chrome-and-glass styling, I find the term "Silver Age" thoroughly suitable. Of course something entirely different is meant by it: the step down from the "Golden Age" , which presumably produced the greatest things. At this point I protest loudly.
The best thing that the Americans have achieved in their spirit of invention is the different designing of the fifties. Cars and rockets came into the picture and blended amiably.
To me, the step after the "Golden Age" is the "Platinum Age." Stylistically, 78 rpm jukeboxes were very influenced by the earliest days. They are heavy and ponderous. How good the new look is, breaking free to stand for progress and the future.
Jürgen: Personally, I can understand that completely, but explain it for us.
Petra: The clear, open and honest appearance of the best pieces in our collecting period fascinates me more from day to day, even though I have lived with such things for many years. The radiance of these objects of music, light and design is enormously positive, an affirmation of life, yes, even happy. In my eyes, these are cult objects.
Jürgen: Let us come back again to the subject of the collector. We have already spoken of what constitutes a collector. Are there, here in Germany, many people who are so involved?

Collectors and Fans

Petra: I think there are comparatively few genuine collectors here in Germany. That is simply because our hobby takes up a lot of space, and for that reason it is impossible for most people.
Of course, we must not forget the incredibly many fans who have one or two boxes in their houses and are happy with them.
Jürgen: And how many are there actually?
Petra: It is hard to say. Naturally, we know a number of fellow collectors personally. Perhaps there are thirty of them. But the tendency is growing steadily.
In international terms, though, Germany is still at the beginning. The Hollanders, for example, are far ahead of us.
Jürgen: You have just said that one needs a lot of space for this hobby. Does one also need a lot of money for it?

Who does not dream of encountering a warehouse where there are really old boxes in good condition? It is important that the entire boxes have remained completely dry. Very often, though, it must be ascertained that the boxes were left somewhere carelessly after their actual use. Damp storage places like barns, garages or cellars have then left their traces that cannot be removed. In this case, the chrome was destroyed, the case was damaged, the works have rusted. Good wares, such as here in a private house in Holland, are always harder to find.

October 15, 1988, 10:00 AM: the first public music box fair in Europe opens its doors at the Autotron in Rosmalen, Holland. More than 1000 juke-box fans formed a line at the ticket booth before the show began.
In all, nearly 20,000 visitors registered on that weekend, including those from Belgium, Germany, Britain, Scandinavia and Switzerland.

Below:
The exhibition halls stuffed full of people to see jukeboxes, here at Rosmalen, Holland, in 1992.

Petra: One cannot get anywhere without it. But when I see how other people own a saddle horse or a sailing yacht, or perhaps afford fast cars, then this passion is no more expensive. The capital put into it, moreover, is not consumed, but simply invested. To that extent, the money is not gone.

Jürgen: What are the most significant pieces from your era? I know you even have private items in your collection. That I find quite remarkable. In the case of some "modern" Seeburg machines, one comes to regard them as sculptures. But start with what you consider most important. What are the most desirable collectors' items?

Petra: Our greatest admiration goes to the American boxes made by AMI, Rock-Ola, Seeburg and Wurlitzer. These firms, particularly in the fifties and early sixties, put a variety of noteworthy boxes on the market. I'll name the best of them in order:

AMI: Model H, I, J, K, Continental and Continental 2.
Rock-Ola: Model 1448, 1452, 1454, Tempo, Tempo II, Regis, Empress, Princess 1493 and Wallbox 1484, plus 451, 453, 464 and 470.
Wurlitzer: Model 1700, 1800, 1900, 2000, 2104, 2100, 2300, 2400 and 2500.

Top Items and Their Manufacturers

Of the German manufacturers, the "Tonomat" by Canteen and the "Fanfare" by NSM deserve mention.

Jürgen: I conclude from this that there are also models that are regarded as not so attractive, perhaps even as ugly, although they have visible playing mechanisms?

Petra: That is how we see it. For example, I might mention the Rock-Ola 1455. The series from which it comes was inherently very successful. But this model was fitted with an ugly imitation-wood apron. It is funny, but this awful-looking machine was widely sold in Germany back then. Presumably it suited the tastes of those times best.

Jürgen: And the previously named models are really good designs?

Petra: Now I have to laugh. This

Who has not, as a child or teenager, admired his idols and celebrated within his own four walls?
Meanwhile, the teenagers grew to be adults, and some of them still love what was modern in their era a quarter of a century ago.
If it concerns music, then jukeboxes surely play the first fiddle.
This young Elvis collector from the Ruhr area of Germany knows what she wants: "Elvis forever."

question was bound to come.
In the sense of "Dieter Rams," surely not. He, whose name is synonymous with good taste in this country, as chief designer of the Braun firm, has really made history. When he says, "Good design is little design," and speaks very clearly of "the gentle order of things," then things are certainly much "louder" here. The music doesn't even need to be playing.
Of course, there were reasons for the glaring appearance of the boxes—animation was the magic word that was supposed to draw the coins out of the pockets. Loud colors, glittering chrome frames and light-filled speaker grilles attracted people like a magnet.
Jürgen: A discovered feast for speculators?

A perfect presentation, already assembled in the eighties, of juke boxes from the fifties by dealer Ben Franse of Holland. His showroom in Scheveningen/The Hague was professionally designed to suit his wares. Meanwhile, the first juke-box dealers in Germany have opened for business.

Attending the first juke-box sales exposition in Germany at Kaunitz, a small town in Eastern Westphalia with a big exposition hall, can make history. Here it is June 26, 1993, and there are exactly thirty people waiting at the door to be the first guests. The promoter, Gerry Mizera of Berlin, called this long-awaited German happening "The Jukin' Fifties."
In just a few years, fairs of this kind will become commonplace in Germany, if developments in neighboring countries reach Germany.

Below:
The jukebox and Rock 'n Roll stick together. They have mutually gained from music.
These four happy contemporaries have known that for a long time. The fifties revival of these times is more than a short walk to nostalgia, it is an outlook on life in every respect.
The splendid AMI I box of 1958 does not leave the true Rock 'n Roller cold. It ranks among the most impressive and often-played jukeboxes with car-like styling, and is thus a particularly sought-after collector's item.

The First Sales Fair in Germany

Petra: In principle, yes. The material, though, is so complicated that laymen are more likely to make fools of themselves than to spot something good.
I always say whoever has too much money should take it to the bank or buy stocks. The "Jukebox Fair" is for the real fans who do everything with their heart.
Jürgen: Such a fair took place for the first time in Germany in June 1993. At that time we looked at the offerings together. For a beginning, that was very good.
Petra: Very much so. It was nice that somebody had taken the trouble to organize such an event. It involved a lot of work and also some risks. For the hall rental, the entertainment program, and particularly the advertising, a five-figure amount had to be drummed up. According to my information, they broke even financially. So it was the first fair, but it won't be the last.
Jürgen: Did you buy a jukebox there, or do insiders get their collectors' pieces elsewhere?
Petra: Since we were just looking for very specific models to complete our collection, we didn't find anything. But there were several specimens among the offerings that were acceptable even to our very high standards of quality.
Jürgen: Does one actually get a better deal from a dealer or privately? The distributors' stocks are thoroughly weeded out by now.
Petra: I can confirm the last for this area, at least as far as the open boxes of the fifties are concerned. There are enough of the closed ones standing around. Many of them come to the fairs now, because there are difficulties in obtaining up-to-date single records now.
Impressive boxes with visible mechanisms can be bought from a dealer or privately as well. The most difficult solution is importing them from America oneself.
Jürgen: Why is that?

America Before the Sell-out

Petra: Transportation. Packing. Paperwork, and also the high cost of personal travel make a reasonable purchase almost impossible.
In any case, at fairs in America they demand hard cash in dollars. I read that regularly in reports.
It becomes just as clear too that the Americans—at least certain ones—now have absolutely no interest in exporting these things on a grand scale. They are probably afraid—and not completely without reason—that one day one might regret having let what was regarded as excess stock be sold to foreigners.
Jürgen: What do you see in the future, then?
Petra: This kind of thinking will keep increasing. I am already predicting that in the long run there will be dealers who will take the goods formerly exported from the USA and sell them back there, because by that time they will realize what they have. This is not so bad, either. Mercedes collectors in Germany are already buying some cars from Florida.
Jürgen: How can you really tell whether a box is worth its price?

The Seeburg V 200 was a masterpiece of appearance and futuristic technology in 1955.
This machine has never lost its fascination. Today it, along with its sister machine, the VL 200, is an absolute "must" in every first-class collection.

Beauty Before Age

Petra: The value of a jukebox is a combination of two components: model and condition. In collectors' circles, a model hit parade has been apparent for a long time already. Just as with oldtime cars, very specific things are very much in demand, others less so. Thus a basic price is created. Just as an example, let's say a box in first-class, playable, clean condition has a value of 6000 marks. The very same model, unrestored, dirty and with bad chrome, and not working, would do well to get 2500 marks.
Jürgen: Many people think the older it is, the higher the price will be.
Petra: That is true only to a certain extent. For example, an AMI I made in 1958 costs three times as much as the three year older AMI G. And small-series production is not necessarily a guarantee for high prices. And to that I am happy to say that it has its reasons. Even then, nobody wanted them.
Jürgen: What are the most expensive boxes from the 45 rpm era?
Petra: The highest prices are paid for two pairs of models, the Seeburg V and VL and the Wurlitzer 2000 and 2100.
Jürgen: And what kind of prices are paid for them?
Petra: In perfect condition, they have long since crossed the ten-thousand-mark barrier, and been followed by several items from the list I mentioned earlier.

Astronauts and rockets were the godparents to this unusual jukebox, the Chantal Meteor. A similar model, much like the one shown here, was originally manufactured in Switzerland. Later the production rights were sold to a British firm.
On account of sales difficulties, the two manufacturers could only produce small series, making this box a treasure for the collector.

Jürgen: What, in your eyes, is the best-looking jukebox?
Petra: At the moment, the Seeburg W. But that changes!
Jürgen: And what is the craziest-looking one?

The fascination of space travel transformed into everyday designs. The AMI Continental II of 1962, with its parabolic mirror observing the ether while the two rubber rockets seem to await lift-off into the heavens.
In the mind of the general public, space travel is seen as progress, as a path into an uncertain but interesting future. Many juke-box manufacturers have taken up this timely theme.

Petra: AMI really created one with its Continentals—and did they ever! Scarcely noticed by a lot of people, there was an equally noteworthy box, first made in Switzerland, then in Britain, the "Chantal Meteor," justifiably a sought-after collectors' item—both rare and expensive.
Jürgen: Would I look in vain at your house for a "Wurlitzer 1015 One More Time," which also plays single records?

Nostalgic Creations Without a Chance

What looks like modern art with three pieces in a row means a lot to the insider. Not much actually agrees in terms of color and original condition. Real collectors find it important that the objects of their longing still have original paint. The chrome must also be original. With increasing collecting activity, examples like these three Seeburg V/VL of 1955/56 can find buyers at any time.

Petra: Nostalgic pieces of this kind might have a high functional value for the basement party room, but as collectors' items, they are, especially in my area, completely uninteresting.
I grant the "German Wurlitzer" all its sales success. Then too, there is an American firm by the name of "Antique Apparatus" that has concentrated on similar goods. Only a short time ago they acquired all the rights to Rock-Ola's onetime empire. The legendary David C. Rockola is said to have swung this deal shortly before his death.
Jürgen: Everywhere new ideas are born to design new "old" jukeboxes. The optical designs for them always reproduce the boxes of the forties, but they are usually equipped with the modern CD technology, a contradiction if there ever was one.
Petra: It may be that there is a market for it. Presumably, one can make money that way. But I want to say very clearly that the history of the jukebox is written, and in fact finished. Everything that happens in that respect today is worthless to our way of thinking; it is nothing.
Jürgen: The first nostalgic box was built by the "American Wurlitzer" firm in 1973, as the "Model 1050." It had a 45 rpm mechanism and was a sales flop at that time, but it has its admirers today.
Petra: It is of no interest to me.
Jürgen: We still need to talk about restoring. First, though, how do you like best to buy boxes? Presumably in original condition? But this concept gets stretched a lot.
Petra: This designation is used today for two completely different conditions of quality.
First, it means unrestored, just as it was discovered. Its condition can be good or bad.
Second, it means an excellent piece, with original chrome, non-repainted case, completely collectable as it is. That's the way I like them best. But they still require a lot of work.
Jürgen: I presume the cleaning cloth now plays its role?
Petra: Precisely. But first comes the screwdriver.
For a thorough cleaning, the box has to be completely dismantled. Naturally, this can be done in stages, so you don't have too many individual parts lying around. It is very important to make sketches and take photos to aid your memory. Every single screw, including those for the bottom plate, should be replaced exactly where it was before. Anything else is a bad job.
Suitable cleaning materials, especially to remove the nicotine deposits that are always found, are called for now.
Jürgen: What kinds do you use?
Petra: Everybody has his own patent recipe. We use an active spray. Panels with writing have to be handled very carefully, because chemicals can make them dissolve easily. Otherwise you do more harm than good.
Jürgen: Can you do all the repair work yourself?
Petra: No, I can't. That's why I have men in the house. Our thirteen-year-old son Jan Paul is very capable, and so is his father. And when they can't do it themselves, they always manage to find someone who can.
Jürgen: You have a quantity of spare parts stored in your cellar.

The good old bass violin, which even accompanied Elvis in the old days, is once again acceptable for Rock 'n Roll bands. The shapeless giant finds merry use in the show as "gymnastic equipment."

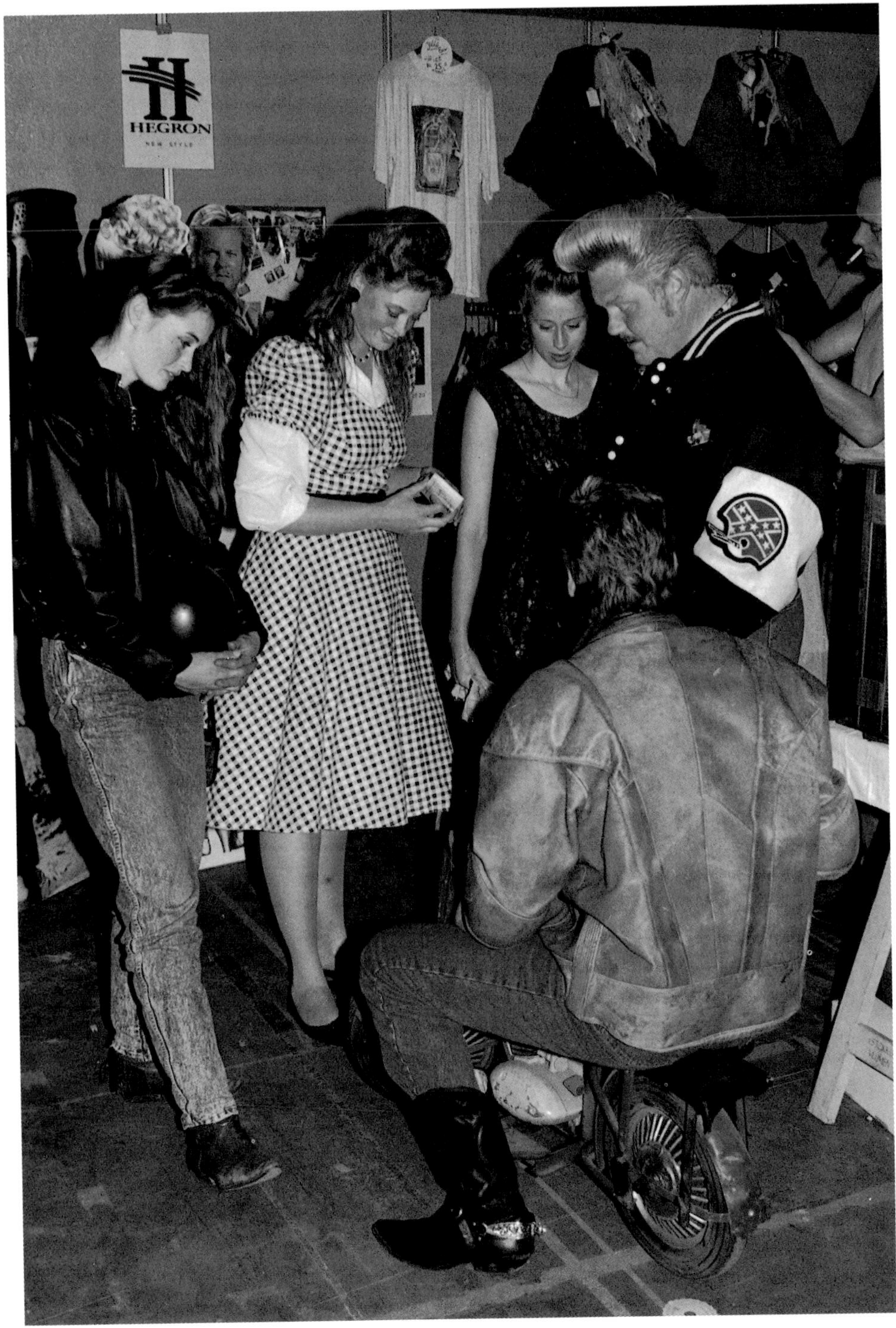

The fifties are being relived by countless young people all over the world. It consists mainly of acquiring expensive industrial products from that era, but music and clothing naturally play a role too.

Lower right:
"My baby, you grew up in a brand new Cadillac," sang the Renegades in 1965. If this girl also goes by the name of Marilyn Monroe, then the evening may be rescued.
The car is genuine, but the Marilyn is not, regardless of how good she looks. But both belong to the scene.

Spare Parts and Restoration

Petra: Really a lot of them. Of course, experience teaches us that exactly the part we need is often lacking. Then we have to contact fellow collectors.
Sometimes we search in vain for years. You have to have a lot of patience.
Jürgen: Can you still get playing systems, needles and tubes?
Petra: Some things are easy to come by, but others hardly turn up any more. If I need a "Momo-red-head-playing system" by Seeburg, then the prospects look very bad. If you can find one at all, it will cost 200 marks. In terms of design, it forms a unit with the tone arm. Any other system would not look right.
Important tubes such as the "6973" are also very hard to get now. Maybe they will be manufactured again some day. I can see a market for them.
Jürgen: The finished tube amplifier certainly has a great sound. Many modern stereo sets cannot match it.
Petra: The best sound quality is found in Seeburg machines. The sound gets under your skin when the right music is played.

Junk is not necessarily junk. The highly specialized collectors like to dig into heaps of individual parts that have no value to the layman.
Restoring a jukebox depends entirely upon the availability of original parts.

Jukeboxes of the "Platinum Age" (1951 to 1964)

Model	Selections	Maker	Month/Year	Notes
E-80	80	AMI	3/1954	
E-120	120	AMI	3/1954	
F-80	80	AMI	1954	
F-120	120	AMI	2/1955	8 diff. colors
G-80	80	AMI	12/1955	
G-200	200	AMI	9/1956	Aquarium/kiosk
H-100	100	AMI	7/1957	
H-120	120	AMI	7/1957	
I-100	200	AMI	9/1958	
I-120	200	AMI	9/1958	
I-200	200	AMI	1958	Keys
I-200 Manual	200	AMI	3/1958	Selection wheel
J-100 Manual	100	AMI	1959	Selection wheel
J-100 Stereo	100	AMI	1959	Keys
J-120 Manual	120	AMI	1959	Selection Wheel
J-120 Stereo	120	AMI	1959	Keys
J-200 E Mono	200	AMI	4/1959	Keys
J-200 E Stereo	200	AMI	3/1959	Keys
J-200 M Mono	200	AMI	4/1959	Selection wheel
J-200 M Stereo	200	AMI	3/1959	Selection wheel
K-100 A Mono	100	AMI	1960	
K-100 A Stereo	100	AMI	1960	
K-120 Mono	120	AMI	1960	
K-120 Stereo	120	AMI	1960	
K-200 A Mono	200	AMI	4/1960	
K-200 A Stereo	200	AMI	4/1960	
K-200 E Mono	200	AMI	1960	
K-200 E Stereo	200	AMI	1960	
K-200 M Mono	200	AMI	1960	Selection wheel
K-200 M Stereo	200	AMI	1960	Selection wheel
Lyric 100 A	100	AMI	9/1960	
Lyric 100 M	100	AMI	9/1960	
Lyric 200 A	200	AMI	9/1960	
Lyric 200 M	200	AMI	9/1960	
Continental A	200	AMI	1961	Keys
Continental M	200	AMI	7/1961	Selection wheel
Continental 2 A	200	AMI	5/1962	Keys
Continental 2 100 A	100	AMI	1962	Keys
Continental 2 A Stereo Round	100	AMI	3/1962	
Continental 2 A Stereo Round	200	AMI	2/1962	
Lyric 2 Stereo Round	100	AMI	1962	
The Minstrel	32	Arcadia	3/1954	British
Tonecolor J 32	100	Automateur	3/1954	Tape; Swedish
Tonecolor J 34	50	Automateur	8/1954	Tape; Swedish
Automatengrammofon	20	Automateur	8/1954	Tape; Swedish
G 80	80	BAL-AMI	3/1957	Balfort; British
G 120	120	BAL-AMI	1/1958	Balfort; British
Junior	40	BAL-AMI	1/1958	Balfort; British
Super 40	40	BAL-AMI	3/1958	Balfort; British

Model	Selections	Maker	Month/Year	Notes
Symphonie	40	Bergmann	4/1954	
Symphonie E 80	80	Bergmann	12/1956	
Symphonie 200	200	Bergmann	3/1958	
Symphonie 120	120	Bergmann	3/1958	Wall box
Symphonie 200 stereo D	200	Bergmann	11/1961	
Harmonie 120	120	Beromat	2/1959	Wall Box
Harmonie 120	120	Beromat	3/1959	Floor box
Harmonie 200	200	Beromat	3/1959	Wall Box
Mignon	56	Bonzini	6/1958	French
Meteor	200	Chantal	12/1958	Flexicoupling
Hit Parade	10	Chic. Coin	1951	Table box
Favorit 60	60	Contina	9/1957	Wall box
Rhythmus	80	Drössler	9/1956	German
Bambi 80	80	Eichhoff	5/1959	German
F 60	60	Eltec	11/1958	Wall box
S 100	100	Eltec	2/1959	Floor box
F 100	100	Eltec	10/1959	Wall box
Europa-Box 4053	40	Europhon	3/1953	German
245 Jubilee	40	Evans	1952	
2045 Century	100	Evans	9/1952	
Holiday	100	Evans	3/1954	
445 Jewel	50	Evans	1954	
100 SEL	100	Evans	10/1958	
Maya 100 stereo	100	Evans	1959	
Magnetic 96	96	Futurity	5/1958	
Disc-o-matic	50	Gama	6/1958	Belgian
2100	50	Gama	10/1958	Belgian
2100	100	Gama	10/1958	Belgian
Favorit 100	100	Hagen-App.	2/1957	German
Phono-Bar		Hecker	3/1958	Cabinet
Phono-Bar "in new colors"		Hecker	1/1959	Cabinet box
Junior	100	Hemann	3/1957	German
J-40	40	IMA-AMI	6/1954	Jensen; Danish
J-120	120	IMA-AMI	6/1954	Jensen; Danish
J-40-G	40	IMA-AMI	1/1957	Jensen; Danish
J-80 G	80	IMA-AMI	2/1957	Jensen; Danish
J-80 H	80	IMA-AMI	1/1958	Jensen; Danish
Miniatur de Lux	60	Kirchhoff	5/1959	German
Selectrophone	120	Mapra	6/1958	French

Model	Selections	Maker	Month/Year	Notes
Emaphone 100	100	Marchant	11/1954	French
Emaphone 100 M	100	Marchant	5/1957	
Emaphone 160	160	Marchant	7/1957	
Emaphone 96	96	Marchant	3/1958	
Emaphone 112	112	Marchant	10/1958	
Emaphone 72	72	Marchant	11/1958	Wall box, 4 legs
Melody Select Eighty	80	Martina, G. 6/1958	French	
M 88	88	Matebois	6/1958	French
S 80	80	Matebois	10/1958	French
Musimatic	96	Musimatic	6/1958	Wall box
Jupiter 100	100	Musimatic	6/1958	French
Jupiter 104	104	Musimatic	5/1959	French
Super Jupimatic 104	104	Musimatic	7/1962	French
Fanfare 60	60	NSM	1/1957	Green
Fanfare 60	60	NSM	4/1957	Red
Fanfare 60 A	60	NSM	10/1957	
Fanfare 100	100	NSM	6/1958	
Fanfare 100	100	NSM	4/1959	3 colors
Fanfare 100 Stereo	100	NSM	9/1959	
CM 80	80	Renotte	6/1958	Belgian
CM 160	160	Renotte	6/1958	Belgian
CM 30	60	Renotte	12/1958	Belgian
W 80	80	Renotte	12/1958	Wall box
Ristaucrat '45'	12	Ristaucrat	10/1950	Table box
Ristaucrat S-45	12	Ristaucrat	9/1951	Table box
1436 Fireball	120	Rock-Ola	4/1953	
1436A Fireball	120	Rock-Ola	1953	
1438 Comet	120	Rock-Ola	12/1953	
1442 Junior	50	Rock-Ola	9/1954	
1446 Hi-Fidelity	120	Rock-Ola	1954	
1448	120	Rock-Ola	9/1955	
1452	50	Rock-Ola	1955	
1454	120	Rock-Ola	6/1956	
1455	200	Rock-Ola	9/1956	
1458	120	Rock-Ola	11/1957	
1462	120	Rock-Ola	3/1958	
1464	120	Rock-Ola	4/1958	Wall box, round disc
1465	200	Rock-Ola	3/1958	
1468 Tempo	120	Rock-Ola	12/1958	Title drum
1468 Tempo Stereo	120	Rock-Ola	12/1958	Title drum
1475 Tempo	200	Rock-Ola	12/1958	Title drum
1475 Tempo Stereo	200	Rock-Ola	12/1958	Title drum
1485 Tempo 2	200	Rock-Ola	12/1959	
1478 Tempo 2	120	Rock-Ola	1/1960	
1484	100	Rock-Ola	10/1960	Wall box, 3 stars
1488 Regis	120	Rock-Ola	10/1960	
1495 Regis	200	Rock-Ola	10/1960	
1494	100	Rock-Ola	1961	Wall box, 3 stars

Model	Selections	Maker	Month/Year	Notes
1496 Empress	120	Rock-Ola	1962	
1497 Empress	200	Rock-Ola	1962	
1493 Princess	100	Rock-Ola	1962	
403	100	Rock-Ola	Wall box	
430	100	Rock-Ola	1965	Wall box, 1 star
M 100 B	100	Seeburg	10/1950	First 45 rpm box
M 100 BL	100	Seeburg	10/1951	
M 100 C	100	Seeburg	5/1952	
100 W	100	Seeburg	8/1953	
HF 100 G	100	Seeburg	9/1953	
HF 100 G de luxe	100	Seeburg	10/1954	
HF 100 R	100	Seeburg	9/1954	Blue cover
100 J	100	Seeburg	8/1955	Red-yellow cover
100 JL	100	Seeburg	10/1956	Red-yellow cover
V 200	200	Seeburg	8/1955	Gray-green cover
VL 200	200	Seeburg	10/1956	Orange cover
KS 200	200	Seeburg	3/1957	3 rear lights
KD 200	200	Seeburg	7/1957	3 rear lights
L 100	100	Seeburg	7/1957	
161	160	Seeburg	4/1958	2 rear lights
161-D	160	Seeburg	10/1958	2 rear lights
161-S	160	Seeburg	1958	2 rear lights
201	200	Seeburg	4/1958	3 rear lights
201-D	200	Seeburg	10/1958	3 rear lights
220	100	Seeburg	10/1958	
222 Stereo S	200	Seeburg	12/1958	
222 Stereo D	200	Seeburg	12/1958	
Fennofen 24		Textoprint	11/1954	Finnish
melodie 80		Tonecolor	7/1957	Belgian
V-102	102	Tonomat	9/1953	
Telematic	100	Tonomat	3/1955	
Telematic	200	Tonomat	2/1957	Dial selection
Panoramic 200	200	Tonomat	2/1959	
Panoramic 200 S	200	Tonomat	2/1959	Stereo
Teleramic	200	Tonomat	1960	Dial selection
Arietta Piccolo	70	Treff	2/1957	Wall box
Melodie	70	Treff	11/1957	Wall box
Melodie (Design Piccolo)	70	Treff	6/1958	Wall box
UMC 100	100	United	3/1957	
UPA 100	100	United	9/1957	
UPA 100	100	United	10/1957	
UPB 100	100	United	1959	Raymond Loewy (design)
UPC 100	100	United	9/1960	Raymond Loewy (design)
UPD 100	100	United	9/1964	Raymond Loewy (design)
Diplomat 40	40	Wiegandt	11/1954	Rebuilt
Diplomat 120	120	Wiegandt	11/1954	
Tonmeister	60	Wiegandt	3/1956	Wall box
Tonmaster	60	Wiegandt	8/1956	Wall box
Diplomat	120	Wiegandt	3/1957	New design
Diplomat C 120	120	Wiegandt	9/1958	First German Stereo

Model	Selections	Maker	Month/Year	Notes
M 40	80	Wiegandt	8/1959	Wall box
Music Mite	10	Williams	1951	Table box
Futurity	25	R. Wolff	7/1954	French
Melodie	Wulff	3/1956		
1250	48	Wurlitzer	1950	78/45/33
1400	48	Wurlitzer	8/1952	78/45/33
1450	48	Wurlitzer	1/1953	78/45/33
1500	104	Wurlitzer	1/1953	78/45/33
1500 A de Luxe	104	Wurlitzer	9/1953	78/45/33
1550	104	Wurlitzer	2/1953	78/45/33
1600	48	Wurlitzer	9/1953	78/45/33
1650	48	Wurlitzer	5/1953	78/45/33
1700	104	Wurlitzer	6/1953	1st carousel box
1800	104	Wurlitzer	4/1955	
1900	104	Wurlitzer	3/1956	
2000	200	Wurlitzer	8/1956	
2100	200	Wurlitzer	1/1957	
2104	104	Wurlitzer	1/1957	
2150	200	Wurlitzer	3/1957	
2200	200	Wurlitzer	5/1958	
2204	104	Wurlitzer	5/1958	
2250	200	Wurlitzer	5/1958	
2300	200	Wurlitzer	2/1959	Mono
2300 S	200	Wurlitzer	5/1959	Stereo
2304	104	Wurlitzer	1959	Mono
2304 S	104	Wurlitzer	5/1959	Stereo
2310	100	Wurlitzer	1959	Mono
2310 S	100	Wurlitzer	1959	Stereo
2400	200	Wurlitzer	5/1960	Mono
2400 S	200	Wurlitzer	5/1960	Stereo
2404	104	Wurlitzer	5/1960	Mono
2404 S	104	Wurlitzer	5/1960	Stereo
2410	100	Wurlitzer	5/1960	Mono
2410 S	100	Wurlitzer	5/1960	Stereo
2500	200	Wurlitzer	1961	Mono
2500 S	200	Wurlitzer	4/1961	Stereo
Lyric (German)		Wurlitzer	5/1961	Wheel, divided
2504	104	Wurlitzer	1961	Mono
2504 S	104	Wurlitzer	1961	Stereo
2510	100	Wurlitzer	1961	Mono
2510 S	100	Wurlitzer	1961	Stereo
2600	200	Wurlitzer	1962	Mono
2600 S	200	Wurlitzer	1962	Stereo
2610	100	Wurlitzer	1962	Mono
2610 S	100	Wurlitzer	1962	Stereo
Finale 120 S	120	6/1959		German
Lytrofon-Box	40	3/1954		
Magnetomat	50	2/1955		
Multiphone		12/1957		
MWA	80	9/1955		Wall box
OSCA	70	6/1959		Wall box, Dutch
Star Box		7/1962		rebuilt AMI
"volkseigene Box"		1/1957		
"volkseigene Box"		10/1959		

Additional Reading

Books

Adams, Frank. *Jukeboxes 1900-1992, Vol. 1.* 1992.
_____. *Rock-Ola Jukeboxes, 1935-1989.* 1989.
_____.*Rowe AMI Jukeboxes, 1927-1991.* 1991.
_____.*Seeburg Jukeboxes, 1927-1989.* 1989.
_____.*Wurlitzer Jukeboxes and other nice things, 1934-1974.* Seattle 1983, 1989.
Bolte, Dieter. *Die Geschichte der Musikboxen.* Dipl.-Arbeit, 1988.
Botts, Rick. *Wurlitzer Jukeboxes.* Des Moines, 1987.
Humphries, Ben C. *The Jukebox Bluebook.* 1990.
Krivine, John. *Jukebox Saturday Night.* The Buckleberry Press, Upper Woolhampton, 1988.
(First edition, New English Library, London, 1977)
Ladwig, Dieter. *Jukebox.*
Linder, Frank Urs. *Swiss Jukebox Art.*
Lynch, Vincent. *American Jukeboxe, the Classic Years.* Berkeley, 1981
_____, and Henkin, Bill. *Jukeboxes, the Golden Age.*
Pearce, Christopher. *Jukebox Art.* H. C. Blossom, London, 1991.
_____.*Vintage Jukeboxes,* Apple Press, London 1988.
Rosendahl, Ger, and Wildschut, Luc. *Jukebox Heaven.* Unterpers b.v., Abcoude, 1991.
Van Grinsven, Johan. *Jukebox Virus,* 1989.

Periodicals

Always Jukin'
221 Yesler Way
Seattle, WA 98104
U.S.A.
Tel. (206) 233-9460,
Fax (206) 233-9871
12 issues per year cost $30 in the USA;
foreign subscriptions cost $40,
or $88 for air mail.

Juke-Joint
Milanstrasse 19
D-13505 Berlin

De Jukebox Fanaat
Elkenburglaan 2
NL-5248 BK Rosmalen

Jukebox Collector Newsletter
Rick Botts
2545 S.E. 60th Court
Des Moines, Iowa 50317
U.S.A.

for EXPORT...

specially reconditioned and refinished late model

PHONOGRAPHS

M-100 B
45 RPM
100 SELECTIO

2122 NORTH WESTERN AVENUE • CHICAGO 47, ILLINOIS, U. S. A.

TOP QUALITY and PERFORMANCE

Our "Constant Operation" Test is the final checking procedure that reveals hidden defects and insures trouble free, profitable operation. All phonographs are tested under conditions simulating normal location operation. This is but one of many ways in which *Atlas protects your investment!*

MODEL M-100 C
45 RPM
100 SELECTIONS

MODEL M-100 A
78 RPM
100 SELECTIONS

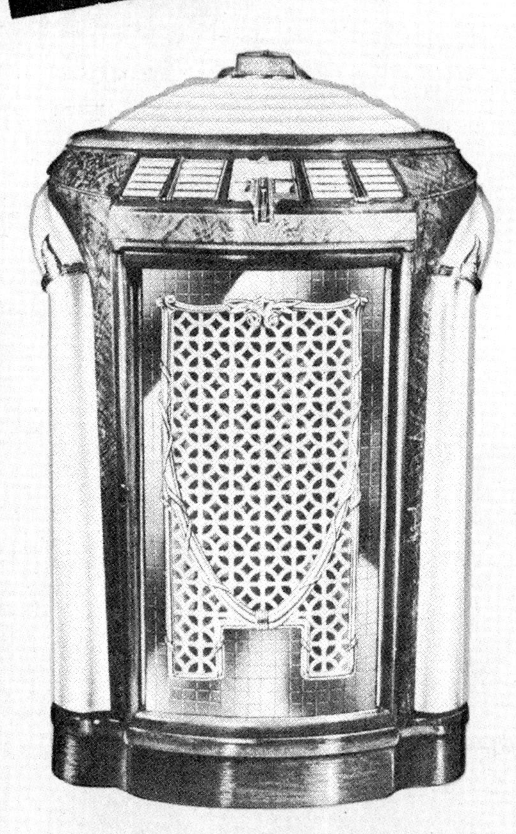

MODEL 1-46 • 1-47 • 1-48
78 RPM — 20 SELECTIONS

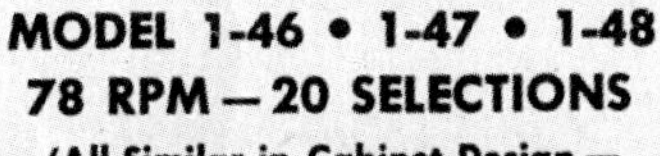

(All Similar in Cabinet Design — can convert to 45 RPM)

All phonographs shown operate on 60 Cycle, 110 Volt AC. 220 Volt AC and 50 Cycles available on all models at additional cost.

ALL PHONOGRAPHS SHOWN ARE

• Atlas Music Company does business with America's leading operators, thus guaranteeing a constant source of late model phonographs and a complete stock suitable for your needs.

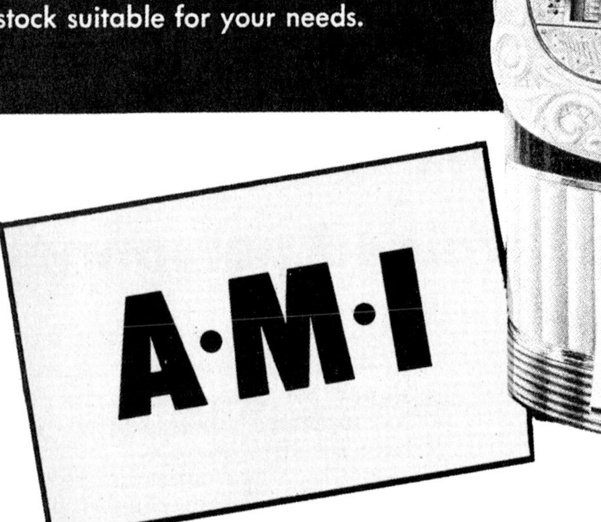

MODEL A
40 SELECTIONS

MODEL B
40 SELECTIONS

Wurl

MODEL 1015
24 SELECTIONS

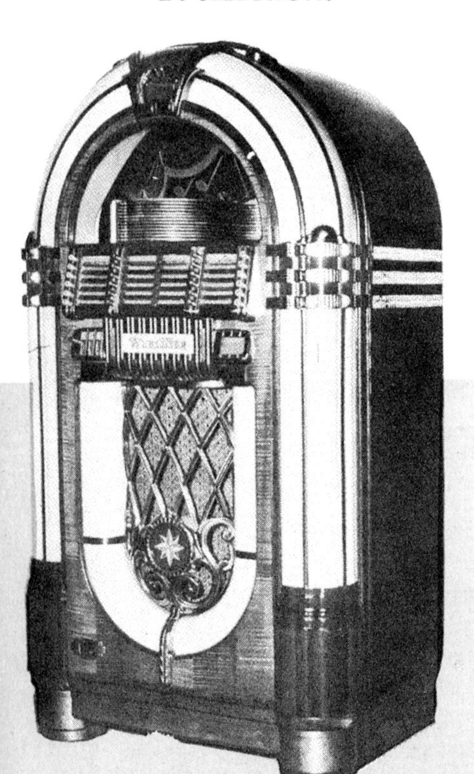

MODEL 1100
24 SELECTIONS

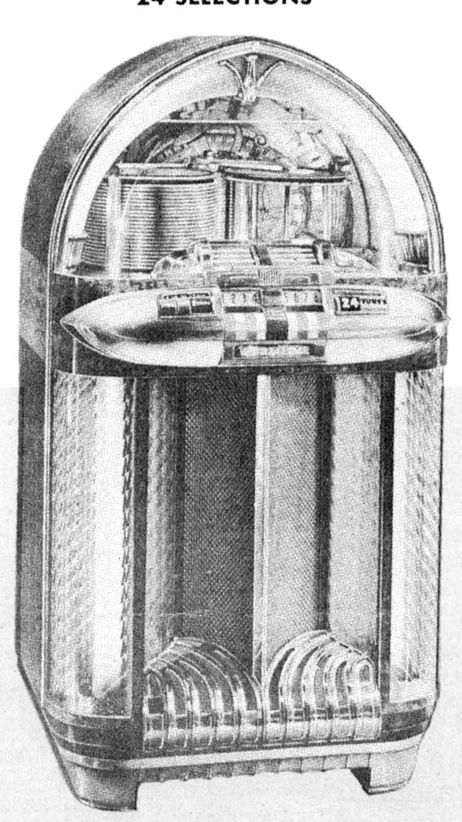

MODEL 1250
48 SELECTIONS

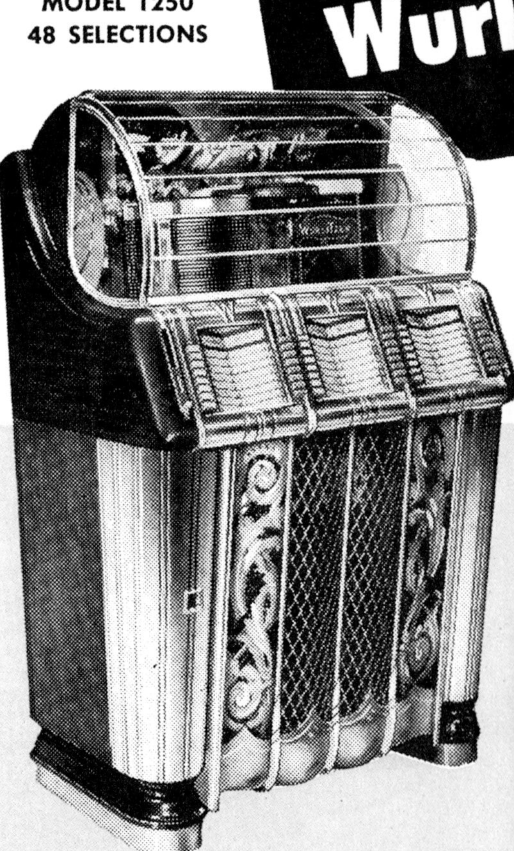

Any Phonograph May Be Fitted

AILABLE FOR PROMPT SHIPMENT

MODEL C
40 SELECTIONS

MODEL D
40 or 80 SELECTIONS

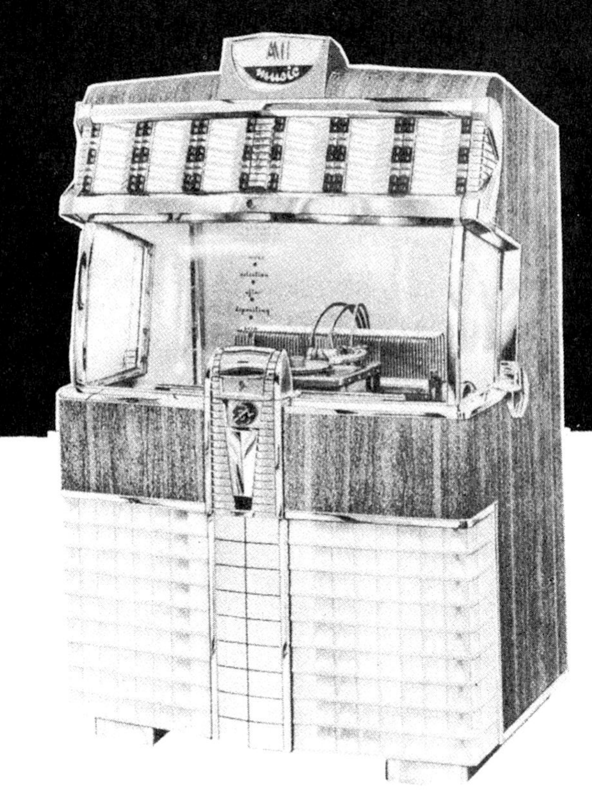

MODEL E
40-80-120 SELECTIONS

MODEL 1400
48 SELECTIONS

MODEL 1500
104 SELECTIONS

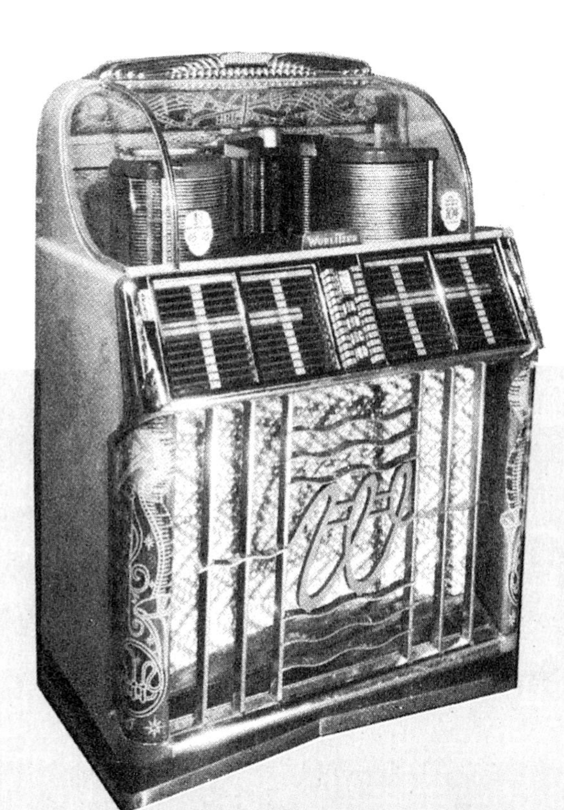

MODEL 1700
104 SELECTIONS

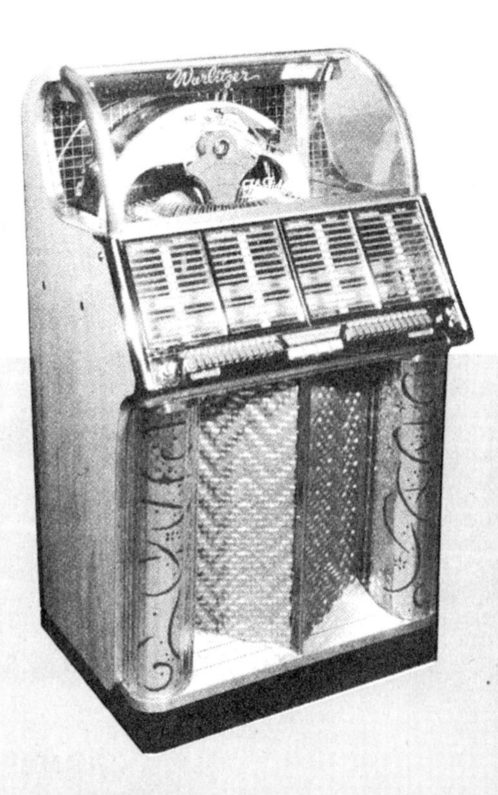

ccept the Coins of Your Country

BUY WITH CONFIDENCE!

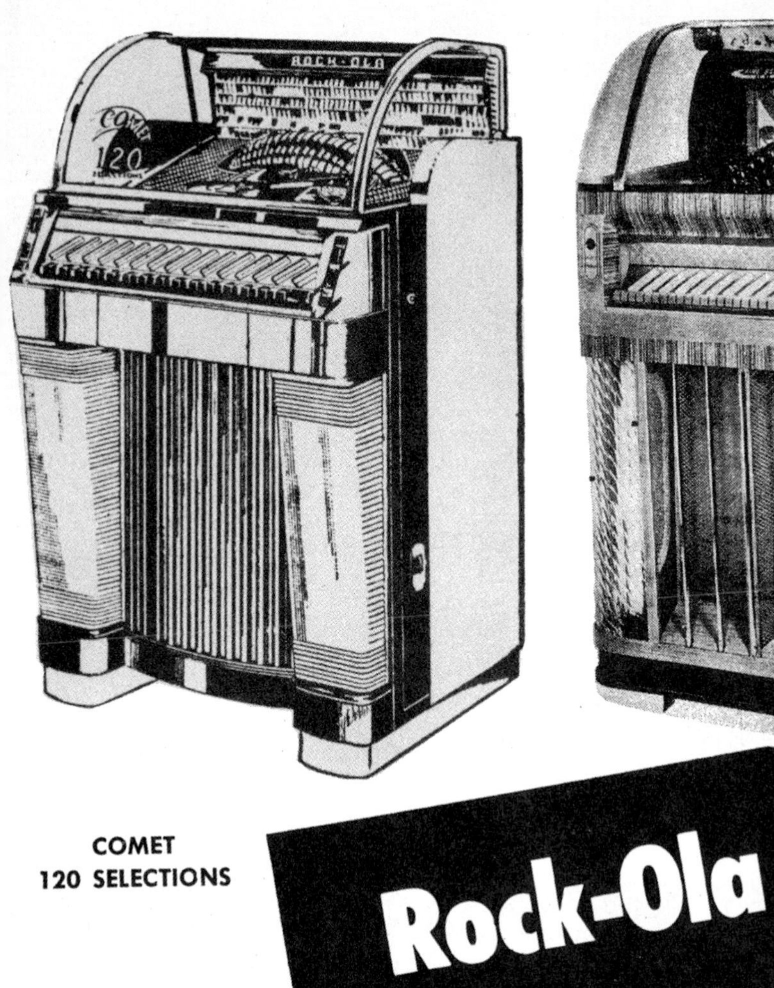

COMET
120 SELECTIONS

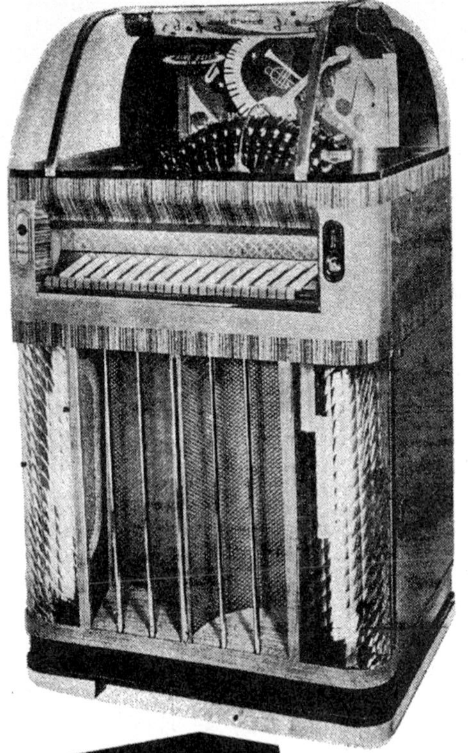

FIREBALL
120 SELECTIONS

MODEL
1422 • 1426 • 142
20 SELECTIONS

YOU CAN DEPEND ON ATLAS!

More than 25 years of experience assure Phonograph Importers of completely satisfactory transactions!

- ***COMPLETE RECONDITIONING***
- ***COMPLETE REFINISHING***
- ***EXPERT PACKING***
- ***FRIENDLY, PERSONAL SERVICE***

ATLAS RECONDITIONING SAVES YOU MONEY . . . includes replacement of worn electrical parts, amplifiers, speakers, etc., with genuine factory units.

ATLAS EXPORT PACKING PROTECTS YOUR PURCHASE . . . special steel-strapped boxes, lined with moisture-proof paper and fitted with interior supports, guard your phonographs all the way.

ATLAS EXPORT ROUTING SAVES YOU TIME . . . so that you may place your phonographs in profitable service at the earliest possible date. Atlas ships via the best, fastest, most economical routes to your country. Your orders will be filled and shipped promptly—you need lose no revenue due to unnecessary delays.

Price Guide

by Richard M. Bueschel

Prices indicated are based on a simple "One Shot" price structure. If you know the key "One Shot" value you can handle just about any buying or selling situation with ease with the knowledge that you are within acceptable pricing ranges. Pricing is based on data provided by auctions, direct sales and dealers in the field.

Here's how the "One Shot" price structure works:

Buying or Selling

The value given is the average going price for a good to excellent condition fully working original and unrestored machine with complete graphics either purchased from or sold by a reputable dealer.

As-Found

Use half of the value indicated for the selling or purchase price from an original owner, picker or estate auction. It is assumed this machine is working but in need of some work and possible restoration.

Basket Case

Use one quarter of value indicated, or even less depending on condition.

Restored Machine

Restorations are generally priced on the basis of the original cost of the machine to the restorer plus the added value of the amount of work put into the machine. As a rule of thumb, a restoration that brought an "As-Found" or a "Basket Case" back to acceptable standards will follow the "One Shot" price, or slightly less. Where the machine has been quickly brought back to acceptable looks with little concern for the original in terms of refinishing to the point of overrestoration the value is somewhat less than that. In the case of an elegant and accurate restoration, preserving the original integrity of the machine and its graphics, the value is a step up from the "One Shot" value.

Price values are in 1996 $US.

Page	Machine	Price
Front cover	Rock-Ola 1454	1300
6 CB	Wurlitzer DEBUTANTE	2200
13 CT	Wurlitzer 1100	4200
18	Seeburg H-100-FG19	1600
20	Wurlitzer 1700	1850
23	Wurlitzer 2504	850
25 TL	Rock-Ola DIAL-A-TUNE	1000
25 TR	Rock-Ola 351	1250
25 BL	Wurlitzer 250 Speaker	450
25 BR	Wurlitzer 340 STROLLER	6500
26	AMI I-200	1800
27	Rock-Ola 1484 Wall Mount	1200
30 CR	Wurlitzer 2150	1350
31	Seeburg V/VL	3400
39 TR	Wiegandt DIPLOMAT	3600
40 CB	Tonomat TELEMATIC 100	1200
41 TL	NSM FANFARE 100 STEREO	1850
41 BR	NSM FANFARE 100	1600
43 TR	AMI CONTINENTAL	2800
54	Wurlitzer 35	2650
56 BL	Wurlitzer 700	4500
56 BR	Wurlitzer 1100	4200
59	Wurlitzer 2000	1800
60 BL	Seeburg 222	1700
62	Wurlitzer 1450	1200
64-66	Rock-Ola 1448	1700
67-69 BL	Seeburg R	2600
70-71	Seeburg G	3200
72-73	Rock-Ola PRINCESS	900
74-76	AMI K	1250
77-79	Wurlitzer 2410	1350
80-81	Rock-Ola 1488 REGIS	1600
82-84	Rock-Ola 1454	1500
85-86	Wurlitzer 1800	3000
87-89	AMI CONTINENTAL	2800
90-91	Tonomat PANORAMIC	2700
92-95	Seeburg 201	2400
96-97	Rock-Ola TEMPO II	1200
98-100	Seeburg VL	3400
101-102	Seeburg J	2400
103 CT	Seeburg WALL-O-MATIC	175
104 TL	Seeburg 3W1	200
104 TR	Seeburg 160	50
104 BL	Hecker-Umbau Seeburg 3W1	250
104 BR	Seeburg 200, V-3WA	225
105 TL	AMI W-120	125
105 TR	Rock-Ola 1555	275
105 BL	AMI WQ-200	150
105 BR	Wurlitzer 5250	150
120 CB	Matezz CHANTAL	6500
120 TR	AMI CONTINENTAL-II	2300
121 TR	Seeburg V/VL	3400
130 TL	Seeburg M100B	1250
131 TL	Seeburg M100C	2500
131 BL	Seeburg 146/147	2600
131 BR	Seeburg M100A	650
132 TC	AMI A	3000
132 TR	AMI B	1200
132 BL	Wurlitzer 1015	9000
132 BC	Wurlitzer 1100	4200
132 BR	Wurlitzer 1250	2000
133 TL	AMI C	900
133 TR	AMI D	750
133 TR	AMI E	1600
133 BL	Wurlitzer 1400	1200
133BC	Wurlitzer 1500	2600
133 BR	Wurlitzer 2304	1450
134 L	Rock-Ola COMET	2500
134 C	Rock-Ola FIREBALL	2000
134 R	Rock-Ola 1422	3500
134 R	Rock-Ola 1426	4500
134 R	Rock-Ola 1428	3000